W9-CHR-559

Robots

Recent Titles in the
CONTEMPORARY WORLD ISSUES
Series

Social Media: A Reference Handbook
Kelli S. Burns

Prisons in the United States: A Reference Handbook
Cyndi Banks

Substance Abuse: A Reference Handbook, second edition
David E. Newton

Campus Sexual Assault: A Reference Handbook
Alison E. Hatch

Sex and Gender: A Reference Handbook
David E. Newton

The Death Penalty: A Reference Handbook
Joseph A. Melusky and Keith A. Pesto

The American Political Party System: A Reference Handbook
Michael C. LeMay

Steroids and Doping in Sports: A Reference Handbook, second edition
David E. Newton

Religious Freedom in America: A Reference Handbook
Michael C. LeMay

Endangered Species: A Reference Handbook
Jan A. Randall

STDs in the United States: A Reference Handbook
David E. Newton

Women in Sports: A Reference Handbook
Maylon Hanold

Books in the **Contemporary World Issues** series address vital issues in today's society such as genetic engineering, pollution, and biodiversity. Written by professional writers, scholars, and nonacademic experts, these books are authoritative, clearly written, up-to-date, and objective. They provide a good starting point for research by high school and college students, scholars, and general readers as well as by legislators, businesspeople, activists, and others.

Each book, carefully organized and easy to use, contains an overview of the subject, a detailed chronology, biographical sketches, facts and data and/or documents and other primary source material, a forum of authoritative perspective essays, annotated lists of print and nonprint resources, and an index.

Readers of books in the Contemporary World Issues series will find the information they need in order to have a better understanding of the social, political, environmental, and economic issues facing the world today.

Robots

A REFERENCE HANDBOOK

David E. Newton

ABC-CLIO™

An Imprint of ABC-CLIO, LLC
Santa Barbara, California • Denver, Colorado

Library of Congress Cataloging-in-Publication Data

Names: Newton, David E., author.
Title: Robots : a reference handbook / David E. Newton.
Description: Santa Barbara, California : ABC-CLIO, [2018] | Series: Contemporary world issues | Includes bibliographical references and index.
Identifiers: LCCN 2018013950 (print) | LCCN 2018018523 (ebook) | ISBN 9781440858628 (ebook) | ISBN 9781440858611 (alk. paper)
Subjects: LCSH: Robots, Industrial—Handbooks, manuals, etc.—Juvenile literature. | Robots—Social aspects—Handbooks, manuals, etc.—Juvenile literature.
Classification: LCC TJ211.2 (ebook) | LCC TJ211.2 .N49 2018 (print) | DDC 629.8/92—dc23
LC record available at https://lccn.loc.gov/2018013950

ISBN: 978-1-4408-5861-1 (print)
        978-1-4408-5862-8 (ebook)

22   21   20   19   18   1   2   3   4   5

This book is also available as an eBook.

ABC-CLIO
An Imprint of ABC-CLIO, LLC

ABC-CLIO, LLC
130 Cremona Drive, P.O. Box 1911
Santa Barbara, California 93116-1911
www.abc-clio.com

This book is printed on acid-free paper ∞

Manufactured in the United States of America

Preface, xiii

1 HISTORY AND BACKGROUND, 3

The Precursors: Automata, 4

How Do Automata Work?, 9

The Transfer of Knowledge: Islam to Europe, 11

Why Automata?, 12

Toward the Golden Age of Automata, 14

Automata in the Modern World, 17

From Automata to Robots, 18

The Birth of the Modern Robot, 20

Evolution of the Robot, 23

    Industrial Robots, 24

    Humanoid Robots, 28

Humanoid Robots in Literature and the Arts, 35

Types of Robots, 38

Conclusion, 43

References, 44

2   PROBLEMS, CONTROVERSIES, AND SOLUTIONS, 61

A Robotic Threat to the Human Species?, 62

Naysayers and Doubting Thomases, 65

Robots and a Jobless Future, 67

    Disruptive Technologies, 68

    Income in a Jobless Future, 71

    Social Benefits in a Jobless Future, 72

    Education and Leisure in a Jobless Future, 74

Moral Robots, 76

Robots at Work Today, 84

    Agriculture, 86

    Business and Finance, 87

    Education, 90

    Comfort Robots, 92

    Health Care, 94

    Military Applications, 99

    Space Research, 106

Learning about Robots, 108

Conclusion, 109

References, 110

3   PERSPECTIVES, 129

Where Are All the Robots? *by Richard Hooper*, 129

Engaging Students through Robotics *by David E. Johnson*, 132

Robotics: A Potential Human Adjunct Needed for the Improvement of Global Health Care and Research Development *by Samuel C. Okpechi*, 136

How My Life Has Been Changed by Robots
by Sierra Repp, 140

Will We Accept Care Robots in Our Homes in
the Near-Future? *by Shalaleh Rismani*, 144

Social Robots for Individuals with Autism
by Anjali A. Sarkar, 147

I-C-MARS Project Built by Native American Students
for Native American Schools and STEM Programs
by Nader Vadiee, 151

Chatbots and Human Conversation *by Erin
Zimmerman*, 156

4   PROFILES, 163

Isaac Asimov (1920–1992), 163

Association for the Advancement of Artificial
Intelligence, 166

Nick Boström (1973–), 167

Rodney Brooks (1954–), 169

George Devol (1912–2011), 171

Joseph Engelberger (1925–2015), 174

Future of Humanity Institute, 177

Future of Life Institute, 178

Heron of Alexandria (ca. 10–70 CE), 180

IEEE Robotics & Automation Society, 181

Institute for Ethics and Emerging Technologies, 184

International Federation of Robotics, 185

Hiroshi Ishiguro (1963–), 187

Al-Jazari (1136–1206), 188

Ray Kurzweil (1948–), 190

Leonardo da Vinci (1452–1519), 193

Machine Intelligence Research Institute, 196

Marvin Minsky (1927–2016), 198

Martin Rees (1942–), 200

Robotics Industries Association, 202

Robotics Institute at Carnegie Mellon University, 205

Victor Scheinman (1942–2016), 206

Alan Turing (1912–1954), 208

Jacques de Vaucanson (1709–1782), 211

5   **DATA AND DOCUMENTS, 215**

Data, 215

Table 5.1. Estimated Yearly Shipments of Multipurpose Industrial Robots in Selected Countries, 215

Table 5.2. Estimated Worldwide Annual Supply of Industrial Robots, 216

Table 5.3. Worldwide Estimated Operational Stock of Industrial Robots, 217

Table 5.4. Use of Robots by Various Industries, 217

Table 5.5. Distribution of Robots among Various Countries and Regions, 218

Documents, 218

Guidelines for Robotic Safety (1999), 218

Efficacy of Robotic Surgery (2013), 224

Idaho Drone Law (2013), 226

Robotically Assisted Surgery (2015), 228

Florida Law on Driverless Vehicles (2016), 231

Draft Report (on Robotics) of the European Parliament (2016), 233

National Robotics Initiative (2011/2016), 236

Preparing for the Future of Artificial Intelligence (2016), 243

Regulatory Robot (CPSC) (2016), 244

Virginia Robot Delivery Laws (2017), 246

Criminal Robots (2017), 247

Taylor v. Intuitive Surgical, Inc. (2017), 249

Amazon Motion to Quash a Search Warrant (2017), 251

Conclusion, 251

6   Resources, 255

Books, 256

Articles, 267

Reports, 279

Internet, 285

7   Chronology, 303

Glossary, 321
Index, 327
About the Author, 339

When I was preparing for the writing of this book, I monitored a number of national and local news sources to see what there was about robots in the news. Hardly a week went by when there were not at least a half dozen articles about some aspect of robotics. Those articles ranged from the somewhat mundane (a robotic device for taking orders, mixing drinks, and serving them to customers at a specialized bar) to much more complex operations, such as assisting in surgical procedures and the creation of original songs and poems. How could a single book deal with the vast number of applications of robots in the world today, let alone the social, ethical, economic, political, and other issues raised by their increasing presence in everyday life?

The answer to that question was, it can't. Indeed, a reader who wants to become really well informed about the science of robotics, its historical development, and its role in modern society should consult a variety of resources. To assist in this research, the author strongly recommends that the reader pay special attention to the references listed at the end of Chapters 1 and 2, as well as the annotated bibliography in Chapter 6. The first of these sources are not only citations to specific references in the text but also guides to further research on many aspects of robotics.

It hardly needs to be said that robots have now become an essential part of the industrial, domestic, research, medical, and other parts of human life. They are used to manipulate heavy machine parts and the smallest components of specialized devices; they travel to environments that are inaccessible or too hazardous for human exploration; they perform surgical procedures with the precision of a human physician; they offer new ways of educating humans; they increase the scope and efficiency of military operations; they make domestic chores easier to perform or even relieve humans from worrying about such chores entirely; they act as aids to many types of scientific research that are repetitious and boring or too dangerous for humans to conduct; and they even act as comforting devices for the aged, young children, and those with medical problems. Indeed, it sometimes seems difficult to mention an aspect of human life in which robots have not or cannot make a significant contribution.

At times, it seems as if robots arrived with the development of modern technologies during the early 20th century. But that is far from the case. Indeed, humans have been building human-like devices for religious purposes, entertainment, military applications, educational purposes, and other reasons since the dawn of human civilization. The earliest oral and written traditions dating back well over 2,000 years contain references to some of these devices, such as the ancient mechanical guardian of Crete in the epic work, *Argonautica*, the Talos. More properly, the Talos was an automaton, a primitive type of robot that fascinated humans until at least the end of the 19th century and continues to do so among many aficionados today.

Chapter 1 of this book provides a somewhat detailed introduction to the evolution of automata over the ages. It discusses examples of the devices that were built, the reasons for their construction, and the effects they had on people of the day. Chapter 2 focuses on the role of robots in today's world, the benefits they have to offer society, the risks they pose, and some of the issues their employment present.

It is important to note that these chapters (and the rest of the book) focus on the development and uses of robots in the Western world, particularly in the United States. In one sense, that is unfortunate since the impact of robots on human society has appeared and promised to expand in virtually all parts of the world, with quite different effects. But the topic of robotics is fundamentally so large and complex that this restriction has been necessary to keep the book of manageable size. The annotated bibliography contains some references to the status of robotics in parts of the world other than the United States.

This book is intended not only to teach about robotics but, perhaps even more important, to act as a tool for young adults who are interested in learning more about the topic and/or want to continue their own research in the field. Remaining chapters of the book, then, provide a review of important individuals and organizations involved with or interested in the field of robotics (Chapter 4), important data and documents dealing with robotics (Chapter 5), an annotated bibliography on the subject (Chapter 6), a chronological history of the development of automata and robots (Chapter 7), and a glossary of important terms in the field. Chapter 3 may be of special interest to readers since it provides the experiences and opinions of individuals who are involved in one or another aspect of the field of robotics.

Robotics is a large and exciting field of study and application. The author hopes that this text provides an overview of the type of progress that has occurred throughout human existence, including the present day; promises of that which is to come in the field; opportunities for young adults interested in the topic; and issues of concern and interest to anyone who wants to know about robots.

Robots

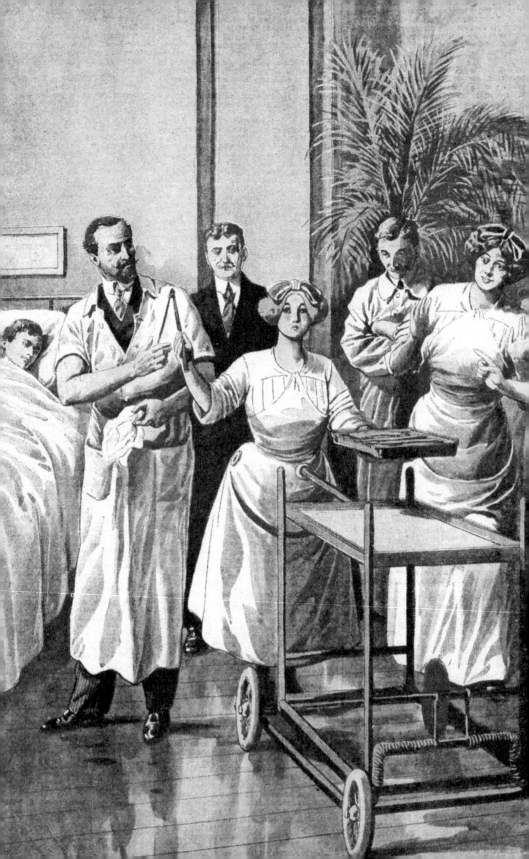

- A shrine over which a bird may be made to revolve and sing by worshipers turning a wheel.
- Figures made to dance by fire on an altar.
- An automaton, the head of which continues attached to the body, after a knife has entered the neck on one side, passed completely through it, and out at the other; the animal will drink immediately after the operation.

These inventions sound like truly remarkable achievements of modern technology: artificial machines—robots—that are able to perform complex actions with little or no input from live humans, except that the machines described here were first built by the inventor Heron (also Hero) of Alexandria sometime before 70 CE (Woodcroft 1851, 93, 95, 109–111).

As the second decade of the 21st century draws to a close, the role of robots in human society appears to be, at the very least, extraordinarily promising. In fact, robots are likely to become so essential to and prevalent in human society that, as one futurist has asked, "the central question of 2025 will be: What are people for in a world that does not need their labor,

---

Mademoiselle Claire, an automaton built by Robert Herdner, is used to hand out surgical instruments from a trolley at l'Hopital Bretonneau in France. From *Le Petit Journal*, Paris, August 18, 1912. (Universal History Archive/Getty Images)

and where only a minority are needed to guide the 'bot-base economy'?" ("Digital Life in 2025" 2014, 11)

Therefore, what has occurred in the field of *robotics* (the branch of technology that deals with the design, construction, operation, and application of robots) between Hero's time and the present day? How has technology changed, and what are the consequences to individuals and society at large of the development of machines that can perform a vast array of "human" tasks? In this chapter, we trace the development of robotics over the past 2,500 years and ask how these developments are likely to impact human society in the coming years.

The Precursors: Automata

The concept of forming inanimate matter, such as clay, stone, mud, and metal, into human-like figures that can perform human-like tasks dates to the earliest days of human civilization. Even before human inventors began to develop the technology for making such figures, legends describing just such objects had appeared. The most general term used for such robot-like figures is *automata* (singular: *automaton*). An automaton can be defined as a mechanism, made of inanimate materials, that performs human-like tasks by following some set of instructions provided by a human.

One of the earliest of these legends is that of the golem, which is probably first mentioned in the Bible (Psalm 139: 16), but that appears more commonly in the Jewish Talmud. The word *golem* can be translated as "shapeless mass" and was used by some early writers to describe the first 12 hours of Adam's life after his creation. The story was that Adam was formed out of a mass of earthy materials that gradually took shape and became a living human. The Talmud and other Jewish documents describe a number of ways in which a golem can be made: by shaping it out of soil and then walking or dancing with it or by writing the name of God on its surface, for example. Later in history, the golem took on other appearances, in some cases, as a kind of monster built to protect Jews from attacks on their communities. (According to modern thinking, the golem may be best thought of as being equivalent to an embryo [Blau, Jacobs, and Eisenstein 2011; "The Golem of Prague" 1948].)

Indeed, tales of statues that could move their hands, feet, and heads; were able to move under their own power; were capable of weeping, sighing, and even speaking; and sometimes were able to make predictions about future events were not uncommon among early writers. One very early work attributed to the Greek god Hermes Trismegistus, for example, claimed that these statues may actually have been imbued with a soul and that their amazing abilities formed the basis of all human religions. (Among the extensive literature on this topic, see Berryman 2009; Brumbaugh 1966; Hersey 2009; Price and Bedini 1964.)

Perhaps the most famous automaton of classical Greece was Talos, a "living" bronze statue of massive dimensions purported to have been forged by Hephaistos (Hephaestus), the god of blacksmiths. Legend had it that Talos was the last of a race of such automata. Talos's task was to patrol the shores of Crete, warning off and destroying forces planning to attack the island. He achieved this goal by throwing massive stones at approaching ships. The automaton was finally defeated through the actions of Medea, the husband of Jason, leader of the Argonauts. Medea was successful against Talos because she was able to remove a plug in the statue that allowed its life-giving fluid, its *ichor*, to flow out of its body, incapacitating the automaton ("Talos" 2017).

Traditional descriptions of automata are hardly reserved to Western civilization. Indeed, they appear to be ubiquitous in most parts of the ancient world. One of the most commonly cited examples of such figures in Indian civilization, for example, is the bhuta vahana yanta (royal mechanical robots), described in the ancient text of Lokapannatti. These figures were mechanical objects armed with swords and stationed at the entrance of a stupa belonging to Ajatasattu, king of Magadha. Any unauthorized person attempting to enter the stupa triggered movement in the figures that caused them to flail their swords at the intruder (Strong 2007, Chapter 5).

Stories of automata in China date back to at least the fourth century BCE. They provide descriptions of rather amazing devices that had the appearance of humans and were able to perform a host of human-like behaviors. One of the earliest of these stories is found in the *Lieh Tzu*, whose author had the same name. He lived during the fourth century and included tales of automata that were made by or for rulers of the kingdom. In one case, for example, King Mu of Chou was introduced to an inventor by the name of Yen Shih, who offered to display for the king a human-like figure he had made. When Yen first appeared before the king with the figure, the king asked, "Who is that man accompanying you?" Yen explained that the figure was an automata, made entirely from leather, wood, glue, and lacquer. At Yen's command, the automaton "walked with rapid strides, moving its head up and down, so that anyone would have taken it for a live human being." Yen then touched the figure's chin, and it "began singing, perfectly in tune." When the king examined the automaton more closely, he found that it also contained perfect replications of a person's internal organs (Needham 1956, 53; for other examples, also see "Highly Advanced Robots in Ancient China" 2015).

The construction and use of automata reached its highest point of development in the ancient world during the peak era of Greek civilization, from ca. 500 BCE to ca. 200 BCE. During that period, a number of inventors—most prominently Archytas of Tarentum (ca. 428–347 BCE), Philon of Byzantium (ca. 280–220 BCE), Ctesibius (ca. 285–222 BCE), and Heron (ca. 10–70 CE)—produced a remarkable number and variety of machines that functioned without human input. The Greeks called these machines *automata*, roughly translated as "things that move apparently by themselves." (Considerable dispute exists as to whether automata are the same as, similar to, or different from modern-day robots. See, e.g., "Difference between 'Robot,' 'Machine,' and 'Automaton'" 2017.) Most of what we know about these inventors and their inventions comes from secondary sources, often written centuries later. Probably the best direct information we have about them is a group of books written by Heron: *Mechanica* (on methods for moving heavy objects), *Pneumatica* (a description of about 80 devices that operate by the use of steam, air, or hydraulic pressure), and *Automata* (explanations of *thaumata* [miracles], such as the automatic opening of temple doors and the dispensing of wine by statues). Perhaps the best example of Heron's work in English is an 1851 translation of *Pneumatica* that includes an introductory treatise on the principles of pneumatics followed by verbal descriptions and diagrams of 78 devices made by Heron (Woodcroft 1851; for a superb review of automata from antiquity to the Renaissance, see Ambrosetti 2010).

The fall of Rome in 476 CE marked the beginning of a 1,000-year era during which interest in secular research declined dramatically. Religious institutions dominated this period and tended to encourage studies relating to church teachings rather than theoretical topics in science and technology. For much of the earlier periods of the Middle Ages, the greatest portion of scientific knowledge available to researchers came from and was based on classical scholars, such as Galen in the field of medicine and Ptolemy in the field of astronomy. Overall, large quantities of learning from the Greek and Roman periods, and even earlier epochs, were lost or, in some cases, systematically destroyed. It is partly for this reason that the period from ca. 500 CE to at least ca. 1500 CE is known not only as the Middle Ages but also, at times, as the Dark Ages (Douglass 2016).

Knowledge about automata faced a similar fate, with many of the original documents from the Graeco-Roman period disappearing forever. Indeed, Western civilization faced the prospect of having to "start over" in most fields of science, mathematics, and technology at the end of the Middle Ages, at least to some extent, were it not for the Islamic civilization that was coming into bloom at about the same time. Following Muhammad's death in 632, the culture he founded began to spread widely throughout the Middle East, northern Africa, and the Iberian Peninsula in Western Europe. The young civilization held high regard for secular learning and soon began to collect and adapt scholarly works from Greece and Rome that would otherwise have been lost forever.

From 750 through 1258, the dominant Abbasid caliphate placed special emphasis on efforts to collect and preserve scientific and technical documents, not only from the Graeco-Roman period but also from civilizations of the East, such as Chinese and Indian cultures. These efforts began when the caliph of Baghdad Abdullah-al-Manum commissioned the Banu Musa ("sons of Musa"), Muhammad, Ahmad, and al-Hasan, to acquire all existing Greek texts for delivery to Baghdad. There the caliph ordered a massive library, Khizanat al-Hikma (The Treasury of Knowledge), and a research institute, Bayt-al-Hikma (House of Wisdom), to be built for the recovered texts. By 1050, the library is said to have collected more than 400,000 volumes from previous cultures, a large number of them original works from ancient Greece. As part of their charge, the Banu Musa also wrote an extensive treatise, *Kitab al-Hiyal* (*The Book of Ingenious Devices*) devoted specifically to a description of automata that they had found. In addition to the "ingenious devices" about which they learned, the Banu Musa described additional devices that they had themselves invented (Nizamoglu 2017).

Probably the best known of all Islamic inventors of the period was the 13th-century Arabic scholar Ibn Ismail Ibn al-Razzaz Al-Jazari (often known just as Al-Jazari). Born in 1136 CE in the city of Jazirat ibn Umar (from which he gets his last name), he gained renown as an artist, craftsman, engineer, and inventor. In 1206, he published a book *Kitab Fi Ma Rifat Al-hiyal Al-handasiyya* (*The Book of Ingenious Mechanical Devices*) that one critic has described as "the most comprehensive and methodical compilation of the most current knowledge about automated devices and mechanics" (Nadarajan 2017). The book contained a comprehensive list and description of a large number of automata known to him, including a variety of time-keeping mechanisms, such as (perhaps most famously) an elephant clock, along with a peacock clock, swordsman clock, and monkey clock; various types of vessels used for drinking; a boat consisting of four musicians who performed automatically for guests at drinking parties; a companion of a king who "drinks the king's leavings"; a slave girl who enters from a closet carrying a goblet of wine; and a variety of devices for raising water (Al-Razzaz al-Jazar 1974; a primary reference for this topic is Zielinski et al. 2015; for illustrations of some devices of the period and their operation, see Dalakov 2017a; "The Elephant Clock by Al-Jazari" 2009; "Pioneers of Engineering: Al-Jazari and the Banu Musa" 2015; Zielinski et al. 2015).

How Do Automata Work?

From the time of their earliest appearance, automata tended to possess a sense of mysticism or magic. Anyone who saw an automaton in operation in ancient China, India, or Greece could hardly avoid imagining that they were observing a sleight of hand performed either by a very clever magician (the inventor) or by some supernatural force (a god or spirit). Thus, the aura created by early automata was commonly associated with religious events or structures (but more about this later) (Ambrosetti 2010; Grafton 2002).

And yet the scientific and technical principles on which automata were based had been known for centuries. The basic structures were almost always one of the six simple machines or some combination of those machines. The structures were then made to move by principles of pneumatics of hydrodynamics that were also well understood at the time (at least by the inventors).

The term *simple machine* refers to one of (usually) six devices that can be used for one of two purposes: increasing or decreasing the force exerted on an object or changing the direction of motion of an object. Those devices are the lever, wheel and axle, pulley, inclined plane, wedge, and screw. Most of these machines were known to early civilizations, although their actions were not expressed mathematically until the Golden Age of Greek civilization. For example, the wheel was probably invented in about 3500 BCE, although its function in transportation devices, which involves the use of an axle, probably did not occur until about 300 years later (Gambino 2009). The use of the lever probably dates to an even earlier era. Historians believe, for example, that Egyptian workers as early as about 5000 BCE were using levers to move heavy objects. It was not until about the third century BCE, however, that inventors made clear scientific and mathematical descriptions of the simple machines. By the end of that period, for example, the inventor Archimedes had described the way in which each of three such devices, the lever, pulley, and screw, worked. In one of the most famous quotations in the history of science, he described what he learned when he said "give me a place to stand and I can move the world [with a lever]" ("Archimedes and the Simple Machines That Moved the World" 2001). An automaton, then, consisted of some combination (a *compound machine*) of these simple machines joined together to move an arm, lift a cup, turn a head, or produce some other action.

The second necessary component of an automaton was some mechanism to provide the force for moving the simple machines. It was of no use just to have a pulley for lifting an automaton's arm if there was no force to make the pulley move. Again, the answer to this problem had been known to humans for centuries: wind and water power. For example, the earliest windmills known were simply devices for catching the force of moving air to move an object (the windmill's fans) to perform some type of work. And, again, this bit of common knowledge was transformed into a scientific principle by a Greek inventor, Ctesibius of Alexandria. In a work he called *Pneumatica*,

Ctesibius described his experiments on the use of air and water to make machines such as an air pump, a water clock, and organs operated by both air and water ("Ctesibius [Ktesibios]" 2008). It was with a combination of these structural components and pneumatic forces that Greek, Chinese, Indian, Islamic, and later inventors were able to produce increasingly complex automata. (For depictions of the mechanisms and means of operation of some automata, see Al-Razzaz al-Jazar 1974; Bur 2016; "The Elephant Clock by Al-Jazari" 2009; Schmidt 2010.)

The Transfer of Knowledge: Islam to Europe

Knowledge of automata, as well as other forms of science and technology, was transmitted from the Islamic empire to Western Europe by way of a number of routes, primarily through the region known as al-Andalus (essentially the Iberian Peninsula), Sicily, and the Byzantium (the region between modern-day Turkey and Eastern Europe), as well as through interactions between Muslims and Christians during the Crusades and commercial and political interactions between the two populations. Possibly the most important of these routes was the first of these, especially at key points in the al-Andalus region, such as Toledo, where the Toledo School of Translators was established by Raymond, archbishop of Toledo, between 1126 and 1151. It provided an important opportunity for Islamic and Western scholars to meet and share information, including much earlier manuscripts from Greece, India, China, Mesopotamia, and other cultures. Similar, less formal, opportunities for exchange were available when soldiers from Christian nations did battle with Islamic armies in Eastern Europe and the Middle East and when scholars and political figures from the West traveled to Islamic states, where they were introduced to the great centers of learning of the time. (One of the best-available discussions of this period is Al-Hassan 2006; also see Douglass 2016; Truitt 2015, 2016.)

Why Automata?

A question in which historians have long been interested is why automata were invented and built in the first place. Probably the four most common answers that have been suggested for this question are (1) as demonstrations of magical processes, (2) for use in religious ceremonies, (3) as toys, and (4) as instructional devices. It seems likely that many individuals at many points in time must have imagined that the moving figures they saw could have possessed this quality only because of some supernatural force unknown to and uncontrollable by human means. In the 12th century, for example, a persistent legend arose that the classic Greek poet Virgil had magical powers that allowed him to build amazing moving devices such as a wooden statue in Rome that would ring a bell when the city was threatened by foreign invades. An attached bronze knight would then point with his spear the direction from which the invaders were coming ("Ancient Temple Inventions Meant to Fool People" 2007; Truitt 2016, 64–67).

A magical aspect of ancient automata seems almost inevitably to be related to their use in religious ceremonies. One possible and perhaps obvious motive would be for priests and other religious leaders to suggest that the movement of machines on their own might actually be a representation of some supernatural force, such as a spirit or a "breath of a god or goddess." The use of automata would, therefore, increase the veracity and power of the religion itself. Indeed, scholarly studies now suggest strongly that automata were built for just such purposes dating back to their earliest construction. Some of the devices that have been described, dating to the end of the fourth century BCE, are

- an automated snail that moved on its own and was able to turn its head, which led a great procession of the Great Dionysia in about 308 BCE;
- a large statue of Nysa, supposedly the nurse of Dionysus, that was able to stand and pour drinks from a flagon into cups at the conclusion of a large religious festival;

- figurines on the altar of a temple that dance when a sacred fire is lit on the altar;
- trumpets that, without the presence of a human player, produce a sound when the doors to a temple are opened;
- a machine that automatically dispenses a fixed amount of sacred water when a coin is inserted into a slot (all examples from Bur 2016).

The appearance and function of such devices in religious ceremonies thus had the ability, according to one scholar, to "invoke a particular type of sacred awe" (Bur 2016, 6; also see "Ancient Temple Inventions Meant to Fool People" 2007; Bosak-Schroeder 2016; some authors question the extent to which automata functioned as religious wonders; see, e.g., Wikander 2008, 785–799).

The claim that some ancient automata were intended strictly as toys is largely speculative. Little direct commentary by the inventors themselves is available to determine why a particular device was made or used. But the very nature of some of those devices would appear to suggest that they had no more intended function than to entertain people, to be used as toys. An expert on the topic has suggested, for example, that a famous steam engine invented by Heron was probably intended only as a toy, with no practical function (Brumbaugh 1966, 4). The same author describes other inventions of ancient Greece that seemed to be designed for no reason other than the entertainment of both children and adults (Brumbaugh 1966, 10, 19, 30, 47, 49, 72, 90, 98, 112, 118; also see Reese 2016).

The notion that ancient automata were built for the purpose of illustrating scientific principles is a somewhat recent one and one that has inspired some controversy among experts in the field. The argument presented is that the golden age in Greece, in particular, was a time during which significant breakthroughs in science and technology were occurring. Some of the individuals responsible for those breakthroughs may, some have argued, have built ingenious mechanical devices to demonstrate these advances. Therefore, Heron's fountain was not just a toy but a practical lesson for the general public as to the power of steam. As one writer has argued, automata "have often been—and occasionally still are—dismissed as worthless toys or 'marvels' intended to evoke religious awe among the superstitious public. For some time, however, a more serious judgment of automata has prevailed, describing them as object lessons in mechanical and pneumatic principles, rather than as tricks intended to inspire wonder" (Wikander 2008, 785).

In conclusion, it is worthwhile to remember that much about ancient automata, including the very reasons for their production and use, is unknown. Original manuscripts have been lost or mistranslated, and reading an inventor's mind at a distance of 2,000 years is, to a large extent, an exercise in a shrewd game of guessing.

Toward the Golden Age of Automata

By whatever mechanism(s) it may have occurred, the transfer of scientific and technological knowledge of ancient Greece and other civilizations had been successfully accomplished by the end of the Middle Ages. Indeed, during that period, inventors continued to imagine and produce ever more complicated, more realistic, and more intriguing devices. In the introduction to her book on automata in the Middle Ages, for example, Bryn Mawr professor E. R. Truitt writes about the period that

> golden birds and beasts, musical fountains, and robotic servants astound and terrify guests. Brass horsemen, gilded buglers, and papier-mâché drummers mark the passage of time. Statues of departed lovers sigh, kiss, and pledge their love. Golden archers and copper knights warn against danger and safeguard borders. Mechanical monkeys, camouflaged in badger pelts, ape human behavior in the midst of a lush estate. Corpses, perfectly preserved by human art, challenge the limits of life. Brazen heads reveal

the future, and a revolving palace mimics the revolution
of the spheres. (Truitt 2016, 1)

The period between ca. 1860 and 1910 is sometimes called
the *golden age of automata*. It was during this period that the
construction of such devices reached their peak of interest to
inventors and popularity among the general public. Companies
across the Western world arose to meet the increasing demand
for automata, commonly because of their appeal as toys. As
one expert has observed, production and sales of automatic
devices during the 50-year period described by the term *golden
age* "thrived as never before" ("The Golden Age of Automata"
2010).

The period between the end of the Middle Ages and the rise
of the golden age of automata (ca. 1500–1900) is marked by
the invention of a number of intriguing devices that continue
to astonish today. For some of these inventions, little other
than the plans for an automaton remains, while, in other cases,
actual models of those devices exist. Perhaps the earliest of these
inventions were those of the great Italian artist and inventor
Leonardo da Vinci. Among the many mechanical devices for
which plans still exist were a mechanical knight and a mechanical lion, the latter designed for King Francis I of France. The
knight was anatomically correct, based on Leonardo's own
extensive studies of the human body. It was to be capable of
sitting down and standing up, waving its arms, moving its head
via a flexible neck, and opening and closing its jaw. The accuracy of the knight's appearance and movement has earned it
from some experts the title of the "world's first robot." The
lion is said to have been able to perform like a live animal and,
at the conclusion of its performance, walk forward, present a
bouquet of flowers, and open its chest to reveal a cluster of lil-
ies (Dalakov 2017b; Taddei 2007; for demonstrations of these
devices, see "Leonardo Da Vinci's Lion Robot for the King of
France, Year-1515" 2008; "Leonardo Da Vinci's World First
Human Robot" 2008).

Some other examples of the automata made during the period between 1500 and 1900 include the following:

- Mid-16th century: The inventor Gianello della tour of Cremona created a number of automata for Emperor Charles V, supposedly to relieve his boredom. One such device was a lute player that could walk in a straight line or in circles while playing her instrument and turning her head from side to side ("Ancient History" 2017).
- About 1615: French engineer Jean Salomon de Caus invented many automata, the most famous of which was probably a device containing birds that chirp and flutter their wings until a mechanical owl approaches them, at which point they stop singing (Dalakov 2017c).
- 1738: French inventor Jacques de Vaucanson invented at least three automata, the best known of which is the Duck, a mechanical device that could eat and drink, swim, digest its food, and produce excreta (video at Goodwin 2010).
- 1846: After a quarter century of research, Austrian inventor Joseph Faber introduced his Euphonia automaton, a machine designed to convert the dots and dashes of telegraphic communication to human speech. The female figure was able to introduce herself to an audience, laugh, speak, whisper, and answer questions from an audience ("Joseph Faber's Amazing Talking Machine of 1845" 2008; Zimmerman and Pratt 2013).

The centers of automaton research during the golden age were Paris and Switzerland. In both regions, a number of small shops appeared for the sale of (usually) small automata to serve and decorative pieces or toys. In many cases, inventors made a conscious effort to increase a person's amazement about a device by making it look ever and ever more human like ("The Magic of Other Countries" 1915, 123–124; Nocks 2008, 37–43).

As an interesting aside, it should be noted that automata were present in a variety of artistic, cultural, and literary events during and after the golden age. Among the best-known examples of this trend were a group of ballets produced in the late 19th and early 20th centuries: "Coppélia" (1870), "Arlequinada" (1900), "Die Puppenfee" (1903), and "Petrouchka" (1911). The first of these, which remains popular, tells the story of a full-size female doll made by Dr. Coppélius that is so life like that one of the villagers in town falls in love with it to the point that he decides to abandon his fiancé and marry Coppélia. The story is based on two earlier short stories, by Prussian writer E. T. A. Hoffmann, *Der Sandmann* (*The Sandman*) and *Die Puppe* (*The Doll*) (Austin 2016). One of today's most popular holiday traditions, "The Nutcracker," also includes a number of dancers representing dolls that have come to life during the production.

In the field of literature, one of the best-known uses of automata occurs in the book *The Invention of Hugo Carbet*, written by Brian Selznick and released in 2007 (Selznick 2007). The book tells the story of the life of French filmmaker Georges Méliès and the numerous automata he used in his motion pictures. The book received the 2008 Caldecott Medal for "the most distinguished American picture book for children." The book was later adapted for motion pictures by Martin Scorsese and released in 2011. It was nominated for 11 Oscars in that year.

Automata in the Modern World

The majority of articles one reads about automata allude to their appearance in ancient times, in the early Islamic civilization, during the golden age of automata, or some other period in the past. The implication appears to be that automata have largely been superseded and replaced by robots. But such is not the case. The invention and construction of automata continues today in the work of an admittedly small but enthusiastic group of men and women. Some sites where more about these inventors can be viewed are those that highlight the work of

Dug North (http://dugnorth.com/), François Junod (http://
www.francoisjunod.com/), Keith Newstead (https://www
.keithnewsteadautomata.com/), Paul Spooner (http://cabaret
.co.uk/artists/paul-spooner/), Carlos Zapata (http://www.car
loszapataautomata.co.uk/), and Bliss Kolb (http://www.bliss
kolbautomata.com/whirligigs.html). (For an overview of the
topic, also see Croft 2017.)

Automata sometimes still occur in other settings also. In
2014, for example, Chinese president Xi Jinping announced
a campaign to promote the teachings of philosopher Wang
Yangming, who lived in the late 15th century. Xi's plan was
to encourage the Chinese people to honor some of the classic
principles on which China was based. One aspect of Xi's campaign was the construction of an automaton very similar to
ancient automata, situated in the city of Guiyang, where people
could come to honor the philosopher. The automaton not only
looks like Wang but is also able to produce calligraphic symbols, which are used to write about 1,000 of Wang's original
aphorisms (Johnson 2017; contains photo of the automaton).

From Automata to Robots

A question frequently asked by students of robots is how these
devices differ from the much earlier devices discussed up to
this point. The answers to that question differ considerably,
depending on the person who is answering the question. In
some cases, the term *robot* (a term sometimes shortened to just
bot) is said to be, for all practical purposes, synonymous with
automaton. According to this answer, robots are only advanced
forms of the automata that have existed for centuries ("Difference between 'Robot,' 'Machine,' and 'Automaton'" 2017).
Other authorities claim that the two terms differ from each
other in everything from trivial to significant ways. For example, robots may be described as more advanced devices that
make use of electronic hardware and computer programming
that was not available to inventors of automata. In addition,
some observers note that automata were built primarily for the

purpose of amazing common people, as toys, magical presentations, or religious symbols, while robots have had a much wider range of practical applications (Kovács et al. 2016).

In any case, it is difficult to say that there was some specific date or period during which the study of automata was *replaced* by a study of robots. Instead, progress in the development of automata/robot technology evolved slowly after the golden age of automata, much as it had done in previous history. New developments made use, of course, of new forms of technology not available to earlier inventors and now fundamental to the operation of human-like mechanical devices ("Automatons" 2013; "ELI5: The Difference between Androids, Automatons, Robots?" 2017).

By the end of the 19th century, a number of developments that set the scene for the creation of devices we now know of as robots had occurred. These inventions came from a variety of fields and usually would not be called robots at all. But they established the technological basis on which the modern robot is based. As just a single example, consider the field of so-called machine tools. A *machine tool* is a device used for cutting, shaping, or otherwise working with metal, wood, or some other material. Primitive types of machine tools existed as far back as 700 BCE, but they became of prime importance in the 18th century, at which point they became a critical contributor to the Industrial Revolution. Some authorities say that the first machine tool was a boring device invented by English inventor John Wilkinson in 1775 (Dinwiddie 2016, Chapter 1). Probably a stronger candidate for a "not-quite-robot" was an invention patented by American inventors Seward Babbitt and Henry Aiken in 1892. That invention was a crane with a "hand" that could grip hot ingots out of a steel-making machine ("1892—Crane—Seward Babbitt (American)" 2017). Some of the other inventions that can be considered as precursors to the development of the modern robot include the following:

1890:   Thomas Edison makes available for sale a "talking doll" that contains a wax cylinder that allows the doll to "speak" for a period of six seconds. The doll was a commercial failure, and production lasted only six weeks. The device was an early effort, however, to provide automatons/robots with the ability to speak like humans (Barajas 2015; contains eight original recordings).

1893:   Canadian inventor George Moore builds a "steam man" that is able to walk on its own, powered by a steam engine inserted into its body. The device smokes a cigar as it moves and looks and walks very much like a human (Hopkins 1897, 377–379; for an automated "electric man" built at about the same time, see "1894–1914—Electric Man—Perew—(American)" 2017).

1898:   Serbian American inventor Nikola Tesla invented a radio-controlled boat whose movement could be directed remotely. In the first demonstration of his invention, Tesla is said to have remarked that "you see there the first of a race of robots, mechanical men which will do the laborious work of the human race" (O'Neill 2012, 191–200).

1912:   American inventors John Hammond Jr. and Benjamin Miessner construct an "electric dog" that moves using the phototropic effect (an electrical system activated by light) (Everett 2015, 408–411; Moffitt 1914).

The Birth of the Modern Robot

One of the most important events in the history of robotics took place at the National Theater of Prague on January 21, 1921. The event was the first performance of a play by Czech playwright Karel Čapek entitled *R. U. R.* The title is an acronym for the phrase *Rossumovi Univerzální Roboti* (*Rossum's Universal Robots*). The play tells the story of the discovery by a marine biologist named Rossum in the future (1932) of a marine organism that behaves like protoplasm. The protoplasm is capable of being converted by chemical means into living animals, including humans. Clever entrepreneurs soon construct a factory for the mass production of these human-like, but synthetic, objects, capable of performing every type of work previously the responsibility of humans. Čapek was stuck for a name by which to call these devices until his brother, Josef, suggested the Czech word *robota*, which translates as "drudgery," "servitude," or "forced labor." (The person so engaged is called a *robotnik*.)

Over time, robots assume all activities once performed by humans, whose life becomes a paradise of leisure-time experiences. Meanwhile a person involved in the manufacture of robots decides that the devices should have a soul in addition to the mental and physical attributes they already possess. That decision turns out to have disastrous consequences (for the humans, in any event) when robots decide to take over the world and eliminate humans entirely. In the play, that decision is announced by a leader of the "robot movement," who exhorts its fellow creatures:

> Robots throughout the world, we command you to kill all mankind. Spare no men. Spare no women. Save factories, railways, machinery, mines, and raw materials. Destroy the rest. Then return to work. Work must not be stopped. ("R. U. R. [Rossum's Universal Robots]" 1920, II-61)

At the end of the play, the robots fare better: two of the most highly advanced (from an emotional standpoint), Primus and Helena, discover that they have fallen in love. They are renamed by the robot leader "Adam" and "Eve"; they join heads and walk off into the sunset with the leader's admonition, "The world is yours" ("R. U. R. [Rossum's Universal Robots]" 1920, E-99; this citation is for the original script of the play).

R. U. R. is a work of extraordinary importance in the history of robots, of course, as the source of the term by which we know such objects today. But it is perhaps even more important because of the vision of "automatons" that the play presents.

These machines are no longer clever technological inventions that perform one or a small number of, admittedly intricate, behaviors. They are now fully automated devices that mimic humans in every way, including the feeling and expression of emotions that appear in the epilogue of the play. As one writer has noted,

> With the advent of Robots, automata become a thing of the past. Automata are unique, hand-created entertainers, whereas Robots are mass-produced workers. (Reilly 2011, 150; we note that Reilly's characterization of automata as a "thing of the past," as noted earlier, is not entirely accurate; for an interesting, but brief, discussion of *R. U. R.*, see "Science Diction: The Origin of the Word 'Robot'" 2011)

R. U. R. debuted on the New York stage in 1922, where it received favorable reviews and ran for 184 performances. (As an interesting aside, two of the robots were played by future film stars Spencer Tracy and Pat O'Brien, both in their Broadway debuts.) The play was never produced as a motion picture, however. Thus, the credit for the first screen appearance of a robot goes to a robotic character known as Marta in the 1927 German-produced silent picture *Metropolis*. In the film, Marta leads a revolt of the poor and working classes against an elite ruling class ("Metropolis" 2017). The robots in both *R. U. R.* and *Metropolis* can all be classified as *humanoids*, devices that have the appearance and at least most of the functions of a human being but that are, in fact, actually machines. In the century that followed, *R. U. R.* inventors continued to make more and more sophisticated changes aimed at both improving the physical appearance of the device and, much more important, increasing and even improving on their ability to carry out, even with greater skill, tasks that are normally conducted by humans.

One of the first humanoid robots built in the early 20th century was called Eric. It was built by English inventors William

H. Richards and A. H. Refell and made its debut at a meeting of the Society of Model Engineers on September 20, 1928. It was, thus, only a year after the first showing of *Metropolis* when the possibility of a humanoid robot was imagined to the construction of such an object. Eric was made of aluminum, weighed about 100 pounds, and looked very much like a medieval knight in armor. The robot stood on and was fixed to a metal box that contained a 12-volt electric motor that sent current through nearly three miles of wiring in its body. Eric toured the world to amazement and praise by both professional scientists and the general public. The robot was eventually dismantled, and some of its parts were used in 1930 to construct an even more sophisticated humanoid robot called George. George, in turn, was taken out of commission and replaced in the early 1950s by an even more advanced model named Robert ("1928—Eric Robot—Capt. Richards & A.H. Reffell (English)" 2017).

Evolution of the Robot

The development of robotic technology after the mid-1920s went in at least two major directions. The first was in the development of larger (or smaller), more sophisticated devices capable of performing tasks difficult or impossible for humans. The robot found along the production line at any car manufacturing plant is a current example of such a robot. Such devices are generally known as *industrial robots*. A good definition for an industrial robot is a "mechanical device that is automatically controlled, versatile enough to be programmed to perform a variety of applications, and re-programmable with a large work space, several degrees of freedom, and the ability to use an arm with different tooling" ("Industrial Robot History" 2017).

The second direction was in the development of humanoid robots that look and act more and more like humans. Such robots are called *androids* (from the Greek *andro-* [man] and *-eides* [like]). Because of the somewhat sexist tone of this definition, the word *gynoids* is also used to describe robots with female characteristics. (The term *fembot* is sometimes also used to refer to a female robot.) Androids and gynoids differ from humanoids, in that the latter has the shape of a human, with a torso, head, two legs, and two arms, but not necessarily the features and actions of a human, while androids and gynoids have all the properties of a humanoid, with the additional attempt to make the device actually look and behave like a human being. Androids and gynoids have appeared in many motion pictures, one of the most recent examples of which was the character Ava in the 2015 film *Ex Machina* (Zuin 2017). Another term commonly used in discussions of human-like robots is *cyborg*. A cyborg (for *cyb*ernetic *org*anism) is a real organism with implanted devices that correct for or extend one's normal functions (Nelson 2013).

Industrial Robots

Credit for the invention of the first industrial robots usually goes to American inventor George Devol. Devol was a prolific inventor, who had worked on the integration of sound into the first talking movies, the world's first automatic door, color printing machines, packaging machines, and, his ultimate achievement, a device he called the Unimate (for universal automation). First installed in 1961 at a General Motors plant in New Jersey, the machine took over the steps carried out by a die-casting machine that was otherwise performed at some risk and not as efficiently by humans. Unimate consisted of a 4,000-pound moveable arm connected to a box containing programmed instructions on a magnetic drum. Advanced versions of the machine were able to carry out other industrial operations, such as welding and moving objects from one place to another (Morrison 2017; Norman 2017; for a contemporary video of the machine's operation, see "Unimate—Robot" 2015).

The basic component of the Unimate was a moveable arm, capable of assuming a number of positions, picking up and manipulating objects, and placing them in some preordained position; it was, that is, similar to a human arm. Much of the history of industrial robots reveals efforts to modify this basic component, allowing it to pick up heavier objects, to manipulate smaller and smaller objects, to make decisions as to where objects are to be placed, and so on. (For a general review of the evolution of robotic arms, see Moran 2007.) In some cases, the research on robotic arms spanned the gap between human-like and industrial applications. One of the first improvements on the robot arm, for example, came in the early 1960s with the invention of the Rancho arm, a device developed by researchers at the Rancho Los Amigos hospital in Downey, California. The arm was developed to assist patients who had experienced damage to their arms. It had six joints, providing it with the flexibility of the human arm. In 1963, the robot arm project was transferred from the hospital to Stanford University, where it went through a number of revisions, all of which were known by some variation of "the Stanford arm." A final model developed in 1969 by then student in mechanical engineering, Victor Scheinman, was the first robotic arm designed to be used exclusively by means of computer control. The arm made use of a computer to carry out fundamental tasks such as the ability to recognize a target, plan for steps in an operation, quantify the force on an object (the sense of "touch"), and manipulate objects by means of some predetermined pattern. Its most significant accomplishments included the assembly of a Ford Model A water pump and partial assembly of a chain saw. By this point, the device had gone beyond its potential use as a prosthetic for humans to a research and industrial robot. It was later sold to the Unimation Corporation, where it became known as the Programmable Universal Machine for Assembly ("Mobie [*sic*] Robot Developed at Stanford" 2017; "The Open Arm v2.0" 2010 [video]).

Inventors explored a number of variations of the fairly simple robot arm exemplified by the Stanford arm. For example, MIT physicist Marvin Minsky developed an arm with 12 joints made to resemble the structure of the crayfish. It was funded by the Office of Naval Research and was designed to be used to manipulate objects underwater. The crayfish model allowed the arm to reach around other objects, pick them up, and move them about (Mitscail 2010 [video]; "1968—Minsky-Bennett Arm—Marvin Minsky and Bill Bennett (American)" 2017).

Some of the earliest research on mobile robots was also under way in the 1960s. For example, as early as 1962, the U.S. Army became interested in the development of a "walking truck" that could be used in settings where trucks, tanks, and tractors would otherwise become stuck. It commissioned General Electric Company to design and build such a vehicle, which became available in 1968. The walking truck was operated by a human seated inside the machine, who controlled its movements by a system of hydraulic valves. The device was also called a *pedipulator* or *cybernetic anthropomorphous machine* ("GE's Walking Truck" 2011 [video]; Mosher 1969).

Another mobile robot of some significance was a robot invented to land on the Moon and travel under its own direction to collect rocks and soil samples. The device, designed by the Space General corporation, had actually been preceded by a Russian device called *Lunokhod 1* (Russian for "moon walker"), sent to the Moon for the Russian mission of the same name (Christy 2017). The Space General device moved on three pairs of legs able to navigate the uneven surface of the Moon. It had one arm with which to pick up rocks and other materials from the lunar surface and a second arm holding a camera with which to examine these materials. Designed to be a part of the 1961 Surveyor mission to the Moon, the device was scrubbed because of weight considerations (Brodsky 2006, 100–102; "Lunar Walker" 2009 [video]). Similar devices were eventually built and used in other space missions. Among the best-known examples of such devices were the Martian land rovers, *Spirit* and *Opportunity*, launched on June 10 and July 7, 2003, respectively. The two robots were expected to continue operation for about 90 days. As it turned out, *Spirit* remained in operation for nearly seven years and *Opportunity* continues, as of early 2018, to travel across the Martian surface collecting rocks, soil samples, and other information about the planet ("Opportunity Update Archive" 2017; "Spirit Update Archive" 2017).

Another trend in research on robots occurred at the University of Edinburgh in the early 1970s. The focus of that research was the use of artificial intelligence (AI) in the design and construction of robots. The term *artificial intelligence* refers to the ability of a machine to perform tasks analogous to those carried out by humans, such as visual perception, speech recognition, decision making, use of one or more languages, and reasoning. The subject of AI is of passing significance in any discussion of robots. As researchers become more and more proficient in the design and construction of androids and gynoids, the emphasis is not so much on making machines that *look* like humans as it is on making them *think* like humans. Indeed, one of the fundamental issues in discussions over modern robots is the extent to which they really do think like humans and not just rely on programs written for them by human inventors.

The first device produced by the Edinburgh researchers was a robot they called Freddy (commonly written as FREDDY) that appeared in increasingly sophisticated forms known as Freddy 1 (1969), Freddy 1.5 (1971), and Freddy 2 (1973). The robot was able to examine a collection of blocks needed to make a simple toy, select the proper sequence for assembling those blocks, and pick them up and place them in the proper location. The simplest example of this work required a minimum of about four hours, largely because of the relatively primitive state of the computers available for guiding the robot. One of the inventors describe Freddy 2 as "probably the most advanced hand-eye system in existence at the time" (Edinburgh Alumni 2015 [video]; Nilsson 2013, 145–147; for a detailed description of Freddy's operation, see Ambler et al. 1975).

In contrast to modifications of robot arms designed to manipulate larger and larger, heavier and heavier objects, a somewhat different line of research involved the development of arms that could identify and manipulate very small objects. One of the first such inventions was the so-called Silver arm, named after its inventor, David Silver. The arm attempted to replicate a human arm that had very sensitive sensors on its "fingertips" that detected small objects and transmitted that information to a computer "brain" ("Timeline of Computer History" 2017; for an interesting contemporary discussion of the technology, see Nevens et al. 1974). Work of this type has eventually led to modern-day arms that are capable of conducting delicate research, industrial, and surgical procedures, even to the point of being able to pick up and manipulate individual molecules (see, e.g., Freitas and Merkle 2002).

Humanoid Robots

Any discussion of the evolution of humanoid robots will involve some overlap with that of industrial robots. Throughout history, many inventors have designed machines that may or may not look like a human but that, in any case, may serve any one of a number of functions, which may include the carrying out of work. Indeed, many industrial robots have been designed to include functions typically associated with human behavior, as moving objects around, recognizing their position in space, arranging them in some desired order, and themselves moving from one place to another. As the previous discussion has shown, a very significant aspect of the development of industrial robots has involved attempts to replicate the motions of the human arm. The history of the Rancho and Stanford arms is a clear example of this truism.

Because of this overlap, historians of robotics tend to disagree as to the device that can be called "the first humanoid robot." Some point to a machine designed (and perhaps built) by Leonardo da Vinci in about 1495. This attribution, however, is based on da Vinci's notebooks that are now more than 500 years old and do not describe an actual working model.

The description da Vinci wrote, however, would match a crude humanoid robot with the ability to stand up and sit down and move its arms, legs, head, and jaw (Moran 2006).

Many of the more advanced automata might also be called the first humanoid robot because they resembled the physical features of a human and were able to conduct one or more human behaviors. The extraordinary robotics Web site, Cybernetic Zoo, for example, lists more than a dozen "early humanoid robots" designed and built prior to 1930. One of the earliest in this list is a "Mr. Eisenbrass" machine, called by its inventor "an automaton." The device had a skin made of rubber and internal "organs" consisting of batteries and platinum wires. Its behaviors were controlled by an organ-type keyboard on which its inventor "played" instructions for standing and sitting, moving its arms and legs, and speaking a limited number of words ("Automaton Extraordinary" 1848).

Any number of devices built in the century following the appearance of Mr. Eisenbrass could be considered to be its technological descendants. Over that time, inventors focused on one or more characteristics a robot needed to have in order to be considered "human like." These characteristics included a relatively small number of human behaviors: the ability to "see," "hear," and "feel" and perhaps also to "smell" and "taste." The device also needed to have the ability to process the information received from these "senses." That is, it needed to know the meaning of each sensory input and how to respond to it. It must, therefore, have some type of brain that would make it able to understand and process the incoming information. Finally, the device needed to have some mechanisms, such as arms, legs, hands, and fingers, for acting on the information received and decisions made by its brain. For example, a humanoid robot might have had to be able to recognize the size, shape, mass, and other properties of some object with which its "fingers" came into contact. Its brain then had to know what it was supposed to do with an object with those properties and to be able to make its appendages respond in an appropriate way (e.g., picking up the object and moving it to some desired position).

Building artificial body parts of a humanoid robot was relatively simple in concept, if difficult in actual practice. After all, scientists had good ideas as to how the human body itself carried out all these activities, using nerves, muscles, and other anatomical and physiological features. An inventor's challenge was to find ways of duplicating those nerves (as with current-conducting wire), the energy required to transmit electrical messages (e.g., with batteries), muscles (as with various combinations of simple and more complex machines), and other body parts.

Probably the most difficult challenge of all, however, was finding a way for a robot to carry out the mental processes needed to receive and interpret information from the machine's sensor and then decide what to do about that information. This field of study, AI, is one of the most exciting and promising fields of robotic research in existence. The story of the history and development of research in the field of AI is extensive and can be discussed only briefly in this book. It is worth saying, however, that the successes of modern humanoid robots depend to a large extent on their ability to "think about" a problem and decide how to act on it.

Advances in the field of AI have led to some extraordinary breakthroughs in recent years. For example, IBM Corporation has been working on an AI device called Deep Blue for more than two decades. The purpose of this research has been to determine if a machine can be built that will demonstrate something that can be called "intelligence" similar to that possessed by humans. One test developed to answer that question has been a chess game between Deep Blue and human chess masters. On February 10, 1996, Deep Blue made the first breakthrough in that regard, defeating chess grandmaster Garry Kasparov in a chess game. The machine lost the match overall, however. Fifteen months later, Deep Blue and Kasparov competed again, with the machine winning the match over-all, 3½ to 2½ (Kasparov 2010). An even more extraordinary breakthrough occurred in 2016 when an AI program called AlphaGo defeated Fan Hui, the European Go champion, and one of the world's greatest competitors in the game of go, often called the most difficult game that exists in the world today, five games to none for Fan (Gibney 2016). The programs available for the construction of "smart" humanoid robots are obviously now available for practical use.

As an aside, one of the most fundamental and perplexing questions in the field of AI is how one knows that a device is actually "thinking." What are the indications that a specific machine actually has developed AI and is processing information the same way that a human does? An answer was proposed to the question by British mathematician Alan Turing in 1950. Basically, Turing said that one could place a device to be tested (e.g., a new humanoid robot) behind a screen and then pose a series of questions to the device. If, Turing said, the device supplied answers to these questions that matched those one would expect from an actual human, then the device could be said to be thinking. Various modifications of this test have been in use in AI research now for more than a half century. (Among the many excellent articles about the Turing test, see "The Turing Test" 1999.)

The first robot to successfully pass the Turing test was a machine called Eugene Goostman, designed and built by Russian inventors. In 2014, the machine took the Turing test and was able to convince a third of the human judges that it could be taken for a 13-year-old human boy 30 percent of the time. The low level of these results was not strong evidence that "Eugene" had passed the test, but many expert observers (but by no means all) argued that it was the first time even this level of achievement had been reached ("No, A 'Supercomputer' Did NOT Pass the Turing Test for the First Time and Everyone Should Know Better" 2014).

As in the discussion of industrial robots earlier, only a small number of the humanoid robots that have been invented can be discussed here. The examples selected are important because of their being pioneer machines of the time or because they contained some type of revolutionary technology.

In the former instance is a robot known as Eric, said to be Great Britain's earliest humanoid robot. Eric was built in 1928 for the opening of the Model Engineers Society convention in London. The opening speaker for the conference was to have been the Duke of York, who canceled his appearance not long before the meeting began. Inventor Alan Reffell decided to solve the problem by inventing an "artificial man," which he called Eric. The robot was made of aluminum and had a voice that was controlled by radio signals from two assistants. At the conference, Eric stood, sat, moved its arms and eyes, and delivered a four-minute opening address ("Eric the Robot Lives!" 2016; Hoggett 2012).

One of the earliest efforts to produce a robot that could approximate human creativity was a device invented in 1931 by American inventor Wycliffe Hill. Sometimes called "the plot robot" or "genie robot," the device was programmed with a series of cards describing the relatively small number of characters and scenarios used in all types of novels, motion pictures, and other literary creations. Hill's book on his invention listed about 30 basic elements involved in the creation of a story that the robot could combine and incorporate in an endless number of ways to produce virtually any book, play, or movie that might be imagined. An aspiring writer could use the robot by spinning a cardboard wheel and note the number that appears in a window. The process was repeated a number of times until some combination of numbers, each represented some specific aspect of the plot, had appeared. Hill published a book on his invention in 1931 that contained the elements of a plot and an insert of the cardboard wheel used in the robot (Hill 1931; Silverberg 2011).

The promise of robots that looked and acted like humans was first broadcast to a very large audience at the 1939–1940 World's Fair in New York City. At that event, the Westinghouse

Electric and Manufacturing Corporation introduced its 260-pound, 7-foot tall humanoid robot, Electro (also called Elektro). The device consisted of 900 electrical and mechanical parts that allowed it to move its limbs and eyes, address the public with a vocabulary of 700 words, and smoke a cigarette ("Electro, the Westinghouse Moto-Man" 2008 [video]; Pierini 2015; "What You Won't See at the World's Fair" 1939).

The story of humanoid robot development since the 1939–1940 World's Fair is largely one of incremental improvements in robot design and operation. An untold number of technological and aesthetic breakthroughs have gradually produced today machines that look strikingly similar to live humans and that are able to perform virtually any human action. In many cases, those developments have come from fields associated with, if not directly considered to be, robot research. For example, efforts to build a humanoid robot have come to rely on better and better understandings by biologists of the components of the human body, such as nerves and muscles, and the way they operate. With this knowledge, robotic engineers are able to create artificial components that more and more closely resemble human body parts and functions themselves.

Since the early 1970s, a significant portion of humanoid robotic research has come from Japan, often under the sponsorship of the Japanese government or large industrial and/or commercial corporations. For example, researchers at the School of Science and Engineering at Waseda University joined together in 1970, to set up a program called The Bioengineering Group. Over the next five decades, the group produced a series of strikingly effective humanoid robots. The first device in that line of research was called WABOT-1 (for WAseda roBOT), released in 1973. The robot was said to have the "intelligence" of a child of one and a half years. It could converse in Japanese, take precise measurements, lift and carry objects, and perform other human-like activities. The robot didn't look very much like a human, but it came close to replicating human activities (Kato et al. 1974; video from this reference at http://rraj.rsj-web.org/en_atcl/159. Accessed on August 14, 2017).

The next product of the Waseda research was WABOT-2, with even more advanced technology. Perhaps of greatest interest was its ability to play the piano and organ. It performed all of the tasks carried out by WABOT-1, in addition to having the ability to read music (through a television camera in its head), play a variety of selections, and even accompany a singer (WABOT-2 2008 [video]; "WABOT-WAseda roBOT" 2003).

One of the interesting phenomena displayed by the WABOT robots was their lack of human characteristics. That is, humanoid robots today for the most part both *look* like and *act* like a human. The most developed humanoid robots today are virtually indistinguishable for humans. Yet throughout the development of humanoid robots, researchers have often sacrificed one of these traits—looks or acts—to produce the other. For example, some of the earliest automata looked strikingly like humans but had very limited human-like functions. In more recent decades, humanoid robots have had vastly improved technology but look little or almost nothing like a human. As an example, the machine that has often been called the first robot to make use of AI, a robot called Shakey, was able to perform some quite remarkable human-like tasks. But it had no resemblance to a human whatsoever (Silva 2011 [video]). Today, most humanoid robots have highly developed technical skills, *and* they look very much like humans. Perhaps the best example of that trend as of late 2017 is a robot developed by Professor Hiroshi Ishiguro of Osaka University. The robot, called Erica, can walk, talk, move its body parts, and move to some extent like a human. And it also looks much like a real Japanese woman ("Talking with a Beautiful Robot Girl" 2015 [video]; also see "Most Realistic Human Robot EVER, Named Sophia Invented by Hanson Robotics" 2016 [video]).

The evolution of humanoid robots is illustrated in research conducted at Honda Company in Japan. In 1986, the company's research division set out on a mission to produce an advanced humanoid robot. Over the next three decades, the company produced a robot (called ASIMO) that was gradually both more technically advanced and more human looking. This progress is depicted in detail at http://world.honda.com/ASIMO/history/.

Humanoid Robots in Literature and the Arts

Automata and robots have been described and discussed in literature and arts from the earliest times. Talos, perhaps the earliest of all automata (discussed earlier), was first mentioned in the Greek epic poem *Argonautica*, dating to about the third century BCE, but possibly based on even earlier writings of Homer (Atsma 2017). In more recent times, one of the earliest use of robots in the arts dates to the production of the French composer Léo Delibes's ballet *Coppélia* (also known as "The Girl with the Enamel Eyes"), which premiered in Paris in 1870. In this work, a life-like automaton (Coppélia) so enchants a young boy in the village (Franz) that he abandons his fiancé (Swanhilda) for the automaton. (All's well that ends well, however, as Swanhilda dresses as the doll and saves Franz from death at the hands of the inventor.) *Coppélia* is still a part of the contemporary ballet canon, and a current research project is being carried out on the development of humanoid robots that will dance, as did the original character in Delibes's ballet (Drake-Brockman 2017).

Humanoid robots are ubiquitous throughout literary history from at least 1900 to the present day. Some examples are as follows:

- Tik-Tok, a character in Frank Baum's 1907 book *Ozma of Oz* (on which the film *The Wizard of Oz* was based). Tik-Tok was a spherical man made of copper whose actions are controlled by a system of clockwork springs. Separate sets of springs operate various body systems, such as thought and action.

- S. Fowler Wright wrote a short story in 1929, "Automata," that mirrored Karel Čapek's play *R. U. R.*, in which a population of robots originally developed to take over the work normally done by humans eventually eliminate the human race.

- In 1964, American polymath Isaac Asimov published a work called *The Rest of the Robots*, consisting of eight short stories and two novels telling of the interaction between intelligent robots and humans. The robots have developed a specialized artificial brain that allows them to communicate on an equal level with humans.

- Writer Robert Mason wrote a novel called *Solo*, in which a robot developed by the U.S. military disappears and is recovered in part by a newer form of the model named Nimroid. The book was later made into a movie of the same name in 1996.

Robots that are able to produce musical sounds had also appeared during the past 50 years, although less commonly than they have in literature. (As noted earlier, however, one of the skills often displayed by automata and robots has been the ability to sing or play an instrument.) One early example of this field is a song entitled "The Robots" written and performed by the German electronic band Kraftwerk in 1978. In that song, singers declare that "We're charging our battery. . . . We are the robots" (music available at https://www.youtube .com/watch?v=VXa9tXcMhXQ).

By far the most common occurrence of humanoid robots in the arts is in motion pictures and television shows. Hundreds of examples of such depictions are available, perhaps the earliest of which was the silent film *A Clever Dummy*, released in 1917. In the *Coppélia*-like film, a man switches places with a robot-like dummy in order to gain the affections of the dummy-maker's daughter (film available at https://archive.org/

details/ACleverDummy). Other, more recent, examples of the appearance of humanoid robots in films and television programs include the following:

- Robbie (or Robby) the Robot, who appeared in a number of films beginning with the 1956 motion picture *Forbidden Planet*. That film was based loosely on Shakespeare's *The Tempest*, in which Robbie corresponds to Ariel, slave to a master named Prospero in the play and Dr. Morbius in the film (film clip at http://www.imdb.com/videoplayer/vi463995161?ref_=nmvi_vi_imdb_2).
- *Westworld* (1973) and *Futureworld* (1976) in which amusement park androids malfunction and begin killing tourists at the parks.
- *The Stepford Wives* (1972), in which the men in the town of Stepford replace their wives with robot duplicates programmed to devote themselves completely to the men's needs and wishes.
- *Blade Runner* (1982), based loosely on Philip K. Dick's science fiction novel *Do Androids Dream of Electric Sheep?*, tells androids indistinguishable from real humans are built to carry out dangerous work on extraterrestrial bodies. The film tells of efforts to find and capture androids that escape into the world of real humans.
- *A. I. Artificial Intelligence*, in which the leading character is an android named David. David has been constructed to duplicate a human boy exactly and then placed in a human family to see how he handles that situation (trailer at https://www.youtube.com/watch?v=oBUAQGwzGk0).
- *Ex Machina* (2015), a highly praised film in which a computer programmer finds himself engaged with a gynoid, Ava, with a very human face but a mechanical body. The affair ends badly for the programmer (trailer at https://www.youtube.com/watch?v=XYGzRB4Pnq8).

One of the intriguing points about the connection between literature and the arts and humanoid robots is recent research that focuses on the invention of such devices that can actually *produce* new music and art. This development goes beyond all previous research because it requires a very high level of AI. This line of research has become possible as a result of advances in *deep neural networks* (*DNN*), a system that involves many layers of stacked data processors (algorithms) that lie between input from sensors and output to actions. Such systems are designed to mimic the behavior of neurons and neural networks in the human brain, thus making a robot capable of activities such as recognizing patterns and making inferences about data.

One of the applications of DNN is the development of robots (usually not humanoid robots) that are fed very large amounts of data, which they use to process and "understand" patterns and possible actions. And one of the outputs of such systems can be devices that can create new patterns of sounds (new music) and colors (art). A research program called Project Magneta has been quite successful in reaching this goal. For examples of the work done by such robots, see Mordvintsev, Olah, and Tyka (2015) and "Analyzing Six Deep Learning Tools for Music Generation" (2017).

Types of Robots

Many types of robots exist today. They have different operating principles, methods of locomotion, practical applications, and other variations. The categories of *industrial robots* and *humanoid robots* discussed so far include a great many specific kinds of robots. But other types of robots that may or may not fit into one of these two categories exist. For this reason, experts have devised other ways of classifying robots beyond the two-category system introduced here. For example, most robots can be classified according to the technical principles on which they are built (Cartesian, cylindrical, spherical, articulated, or Selective Compliance Assembly Robot Arm) ("Types of Robots" 2013).

A discussion of these categories requires somewhat more technical details than can be provided here. Another system of classification is based on the mechanism by which a robot moves. This method of categorization would include the following classes: stationary robots, wheeled robots, legged robots, swimming robots, flying robots, rolling robots, swarm robots, modular robots, micro robots, nano robots, and soft/elastic robots. Each of these categories can, in turn, be further subdivided. For example, wheeled robots can be further classified as single-wheeled (ball), two-wheeled, three-wheeled, four-wheeled, multi-wheeled, or tracked robots ("All Types of Robots—By Locomotion" 2016; one of the best textual and visual discussions of robot types can be found in "World Robotics 2017" 2017; see chapter 1 at https://ifr.org/downloads/press/02_2016/WR_Industrial_Robots_2016_Chapter_1.pdf).

Perhaps the most interesting and useful method of classification for readers of this book is based on the practical applications of a robot. The following robot applications according to this method of classification are domestic, medical, service, military, space, entertainment, and hobby and competition. Again, the vast number and variety of specific robots make a detailed description of such devices beyond the scope of this book. The following list may provide a flavor, however, of the types of machines that are available in each of these categories. (For a good general overview of this topic, see "A Roadmap for US Robotics: From Internet to Robotics" 2016.)

- Having some type of machine, such as an android, to take care of the many routine tasks needed to keep a home functioning has been a fantasy for many inventors throughout the ages. Today, such devices are either currently available or in advanced stages of development. Domestic robots are designed to carry out many such tasks both inside and outside the home. Such tasks may include dusting and vacuuming, mopping the kitchen floor, cleaning a grill, ironing clothes, cleaning the cat litter box, and monitoring and, to a limited extent, caring for babies. Outdoor tasks may include watering the lawn, tending to the garden, and cleaning a gutter. Domestic robots also have other types of tasks, such as providing objects with which children and adults can play (e.g., robotic dogs, cats, and other animals) and acting as social companions for young children, the elderly, or the disabled (video demonstration at "Top 3 Incredible Home Robots" 2016). One of the interesting points about domestic robots is that they may very well not take the physical form in which humans have long envisioned them: an android that looks like a butler or that looks like a human maid, for example. Many of the most popular domestic robots available today—and in the process of development—are, instead, rather mundane-looking cylinders, spheres, cubes, or other geometric figures. They perform many of the household duties described earlier, but they look nothing like a human worker (Carr 2017).

- Robots have now been designed and built for a variety of medical applications, including the drawing of blood; for use in moving hospital equipment and patients; for conducting medical tests such as colonoscopies; as exoskeletons for patients with spinal cord injuries; as aid workers in senior and nursing home care facilities; as mechanisms for assisting in long-range diagnosis of medical conditions; as medical intake processors; and even as comfort "animals" for comforting agents for patients. Perhaps the most promising and most controversial applications are in the field of surgery. Today robots are available, or are in the process of development, which can carry out many types of surgeries normally performed by human surgeons, such as coronary artery bypass, gallbladder removal, excision of cancerous tissue, hip replacement, hysterectomies, kidney removal, kidney transplant, mitral valve repair, pyeloplasty, pyloroplasty, radical prostatectomy, radical cystectomy, and tubal ligation ("Types of Procedures" 2017; also see http://spectrum.ieee.org/video/robotics/medical-robots/

robot-surgeons-are-taking-over-the-operating-room for videos of robotic surgical procedures).

- Service robots may be classified into one of two categories: domestic robots, of the type described earlier, and industrial robots that carry out tasks in environments that are otherwise inaccessible to or hazardous for human activities. Among the settings in which an industrial service robot might be put to use are situations in which levels of nuclear radiation or high temperatures are too risky for humans to enter. The search for and disabling of land mines is another setting in which robots can be used. Robots have also been developed to examine deep mines, underwater environments, and industrial infrastructure that are difficult or impossible for humans to reach, such as underground pipelines or very high structures. Much of the work done in cleaning up after the Chernobyl nuclear power plant disaster of 1986 was done by robots ("Robot inside Chernobyls [sic] Sarcophagus" 2009 [video]). One of the experimental industrial service robots that have been developed is Tactical Hazardous Operations Robot, whose actions are illustrated at https://www.youtube.com/watch?v=LvzMq1QHhwo. Perhaps one of the most advanced, intriguing, and controversial lines of robotic research is the "personal companion" or "erotic" robot, which is attempting to produce androids and gynoids with whom a person can have sexual relationships. As one might expect, that line of research is the subject of intense debate (Kleeman 2017).

- The extensive number and variety of threats to national security and the highly developed character of modern warfare have resulted in the development of a host of robotic systems for dealing with these issues. Perhaps the most widely used and best known (to the general public) of these systems are drones, unmanned flying devices designed to reconnoiter, report on, and attack targets of interest to a military unit. The details of drone operations are beyond the scope of this book, but see Thompson (2016). An example of a robot-controlled weapon system is the Dutch system called Goalkeeper CIWS, first introduced in 1979. The system performs every step in a search-and-destroy program, from surveillance and detection to destruction to selection of the next priority target ("Goalkeeper CIWS Gun System" 2006 [video]).

- Space research is possibly the most obvious setting in which robots can be used, have been used, and will continue to be used in the future. Exploration of the Moon and Mars by humans is certainly very difficult, and visits to more distant bodies by humans will probably be impossible for many years or decades, if ever. For more than half a century, then, researchers have been designing and building devices that can travel into the harsh conditions of outer space; land on the Moon, Mars, and other solar bodies; move about within these hazardous environments; and report back to Earth about their findings. As an example of the many uses to which space robots can be put, the Japanese Space Agency in 2014 launched into space the first talking robot. The robot was unusual in that it could respond to questions not with prerecorded responses but with responses that it produced from its own internal vocabulary of words (Çaliskan 2014; Chemistry Columbia 2012).

- Entertainment robots have been around for decades. Arguably, the leader in developing such robots has been the Walt Disney Company, which began producing humanoid and animal-like robots in the 1960s. The first major demonstration of these devices took place at the 1964 World's Fair in New York City. One of the "stars" of that event was an Abraham Lincoln talking robot that spoke to the audience about the great ideas of American history. Since that time, Disney has produced almost every imaginable life-like robot, from imaginary characters from outer space to a variety of prehistoric animals (Taylor 2015). Today, the types of entertainment robots available to the general public have broadened dramatically and include robotic pets and toys for home use, "comfort" robots designed to interact in a personal way with humans, and professional entertainment robots, such as those that might perform on stage (see, e.g., Lin et al. 2012).

- An increasingly popular application of robots involves their use as hobbies or participants in educational competitions. Such activities are powerful tools in helping individuals, often children or young adults, to learn more about the manufacture of a robotic device and its use in some particular application. One striking example of this field is a program known as FIRST (for "For Inspiration and Recognition of Science and Technology") Global. FIRST Global was originally imagined in 1989 by inventor and entrepreneur Dean Kamen to encourage interest in robotics and active participation in their construction and use. The first FIRST Global Competition was held in Washington, D.C., in July 2017. Teams from 163 nations participated in the games, with each team consisting of anywhere from three to seven members between the ages of 15 and 18. The challenge in the competition was to develop robots capable of distinguishing between plastic balls of two different colors, representing "clean water" and "impure water," as an introduction to problems of water purity throughout the world (FIRST Global 2017).

Conclusion

Humans have been interested in the concept of a robot for thousands of years. The appeal of a seemingly magical device that looks like a live human (or a dog, a cat, or some other animal) has led inventors to create many robotic devices, ranging from the simplest automatons of ancient history to the machines currently available that use the best-available programs for AI. These devices have had a host of applications, ranging from religious festivals and popular entertainment to research in space and human companionship. But developments in robotics have always, and almost inevitably, created a number of nontechnical issues for the general society. What is the risk, for example, that androids might eventually become capable of taking over human tasks for which they were not really intended? Questions such as this one will be the subject of Chapter 2 of this book.

References

Al-Hassan, Ahmed Y. 2006. "Transfer of Islamic Science to the West." Foundation for Science, Technology, and Civilisation. http://www.muslimheritage.com/uploads/Transfer_of_Islamic_Technology_to_the_West.pdf. Accessed on July 25, 2017.

"All Types of Robots—By Locomotion." 2016. Robotpark.com. http://www.robotpark.com/All-Types-Of-Robots. Accessed on August 8, 2017.

Al-Razzaz al-Jazar, Ibn. 1974. *The Book of Knowledge of Ingenious Mechanical Devices*. Translated and annotated by Donald Routledge Hill. Dordrecht: D. Reidel. https://marcell.memoryoftheworld.org/Ibn%20al-Razzaz%20al-Jazari/The%20Book%20of%20Knowledge%20of%20Ingenious%20Mechanical%20Devices%20(2668)/The%20Book%20of%20Knowledge%20of%20Ingenious%20Mechani%20-%20Ibn%20al-Razzaz%20al-Jazari.pdf. Accessed on July 24, 2017.

Ambler, A. P., et al. 1975. "A Versatile System for Computer-Controlled Assembly." *Artificial Intelligence* 6(2): 129–156. http://groups.inf.ed.ac.uk/vision/ROBOTICS/FREDDY/Barrow/Documents/1975_ambler_barrow_et_al_versatile_assembly_AIJ.pdf. Accessed on August 11, 2017.

Ambrosetti, Nadia. 2010. "Cultural Roots of Technology: An Interdisciplinary Study of Automated Systems from the

Antiquity to the Renaissance." https://air.unimi.it/retrieve/handle/2434/155494/134586/phd_unimi_R07642.pdf. Accessed on July 25, 2017.

"Analyzing Six Deep Learning Tools for Music Generation." 2017. The Asimov Institute. http://www.asimovinstitute.org/analyzing-deep-learning-tools-music/. Accessed on July 16, 2017.

"Ancient History." 2017. Automata. http://www.mechanical-toys.com/History%20page.htm. Accessed on August 1, 2017.

"Ancient Temple Inventions Meant to Fool People." 2007. YouTube. https://www.youtube.com/watch?v=r7BHvN 6rZZA. Accessed on July 27, 2017.

"Animation of Al-Jazari's Elephant Clock." 2018. YouTube. https://www.youtube.com/watch?v=MCW_wp0dgF4. Accessed on March 15, 2018.

"Archimedes and the Simple Machines That Moved the World." 2001. Encyclopedia.com. https://www.encyclopedia.com/science/encyclopedias-almanacs-transcripts-and-maps/archimedes-and-simple-machines-moved-world. Accessed on March 15, 2018.

Atsma, Aaron J. 2017. "Talos." http://www.theoi.com/Gigante/GiganteTalos.html. Accessed on August 15, 2017.

Austin, Linda M. 2016. "Elaborations of the Machine: The Automata Ballets." *Modernism/Modernity* 23(1): 65–87. https://muse.jhu.edu/article/609693. Accessed on August 3, 2017.

"Automaton Extraordinary." 1848. *Scientific American* 3(4): 272. https://ia600607.us.archive.org/12/items/scientific-american-1848-05-13/scientific-american-v03-n34-1848-05-13.pdf. Accessed on August 12, 2017.

"Automatons." 2013. All on Robots. http://www.allonrobots.com/automatons.html. Accessed on August 2, 2017.

Barajas, Joshua. 2015. "Child's Toy or the Voice of Nightmares? Listen to the Edison Talking Doll." The

Rundown. PBS NewsHour. http://www.pbs.org/newshour/rundown/creepy-edison-dolls/. Accessed on August 4, 2017.

Berryman, Sylvia. 2009. *The Mechanical Hypothesis in Ancient Greek Natural Philosophy*. New York: Cambridge University Press.

Blau, Ludwig, Joseph Jacobs, and Judah David Eisenstein. 2011. "Golem." Jewish Encyclopedia. http://www.jewishencyclopedia.com/articles/6777-golem. Accessed on July 17, 2017.

Bosak-Schroeder, Clara. 2016. "The Religious Life of Greek Automata." *Archiv für Religionsgeschichte* 17(1): 123–128. https://www.degruyter.com/view/j/afgs.2016.17.issue-1/arege-2015-0007/arege-2015-0007.xml. Accessed on July 28, 2017.

Brodsky, Robert F. 2006. *On the Cutting Edge: Tales of a Cold War Engineer at the Dawn of the Nuclear, Guided Missile, Computer and Space Ages*. New York: Gordian Knot Books.

Brumbaugh, Robert S. 1966. *Ancient Greek Gadgets and Machines*. New York: Thomas Y. Crowell. Available through Open Library (free registration) at https://openlibrary.org/books/OL5993983M/Ancient_Greek_gadgets_and_machines. Accessed on July 18, 2017.

Bur, Tatiana. 2016. "Mechanical Miracles: Automata in Ancient Greek Religion." University of Sydney. https://ses.library.usyd.edu.au/bitstream/2123/15398/1/bur_tcd_thesis.pdf. Accessed on July 24, 2017.

Çaliskan, Murat. 2014. "First Humanoid Robot in Space Chats with Astronaut." YouTube. https://www.youtube.com/watch?v=OXiX2vNe6s8. Accessed on August 21, 2017.

Carr, Nicholas. 2017. "These Are Not the Robots We Were Promised." *New York Times*. https://www.nytimes.com/2017/09/09/opinion/sunday/household-robots-alexa-homepod.html?mcubz=3&_r=0. Accessed on September 11, 2017.

Chemistry Columbia. 2012. "The Space Robotics." YouTube. https://www.youtube.com/watch?v=6cji_huFvrw. Accessed on August 21, 2017.

Christy, Robert. 2017. "The Mission of Luna 17." http://www.zarya.info/Diaries/Luna/Luna17.php. Accessed on August 10, 2017.

Croft, Michael. 2017. "Contemporary Automata." http://www.contemporaryautomata.com/index.html. Accessed on August 3, 2017.

"Ctesibius (Ktesibios)." 2008. Encyclopedia.com. http://www.encyclopedia.com/people/science-and-technology/technology-biographies/ctesibius. Accessed on July 25, 2017.

Dalakov, Georgi. 2017a. "The Arabic Automata." History of Computers. http://history-computer.com/Dreamers/Arabic.html. Accessed on July 19, 2017.

Dalakov, Georgi. 2017b. "The Automata of Leonardo da Vinci." History of Computers. http://history-computer.com/Dreamers/LeonardoAutomata.html. Accessed on July 31, 2017.

Dalakov, Georgi. 2017c. "Salamon de Caus." History of Computers. http://history-computer.com/Dreamers/Salomon_de_Caus.html. Accessed on August 1, 2017.

"Difference between 'Robot,' 'Machine,' and 'Automaton.'" 2017. English Language & Usage. https://english.stackexchange.com/questions/70384/difference-between-robot-machine-and-automaton. Accessed on July 22, 2017.

"Digital Life in 2025: AI, Robotics, and the Future of Jobs." 2014. Pew Research Center. http://www.pewinternet.org/2014/08/06/future-of-jobs/. Accessed on July 17, 2017.

Dinwiddie, Keith. 2016. *Basic Robotics*. Boston, MA: Cengage Learning.

Douglass, Susan. 2016. "Legacies and Transfers: The Story of the Transfer of Knowledge from Islamic Spain to Europe."

http://www.islamicspain.tv/For-Teachers/11_Legacies%20
and%20Transfers%20Story%20of%20Transfer%20of%20
Knowledge.pdf. Accessed on July 24, 2017.

Drake-Brockman, Geoffrey. 2017. "The Coppelia Project."
http://www.coppeliaproject.com/. Accessed on August 16,
2017.

Edinburgh Alumni. 2015. "Freddy II." YouTube. https://
www.youtube.com/watch?v=SJ0qAcpklWQ. Accessed
on August 11, 2017.

"1894–1914—Electric Man—Perew—(American)." 2017.
Cyberneticzoo.com. http://cyberneticzoo.com/walking-
machines/1894-1914-electric-man-perew-american/.
Accessed on August 4, 2017.

"1892—Crane—Seward Babbitt (American)." 2017. Cyber
neticzoo.com. http://cyberneticzoo.com/not-quite-robots/
1892-crane-seward-babbitt-american/. Accessed on
August 3, 2017.

"Electro, the Westinghouse Moto-Man." 2008. YouTube.
https://www.youtube.com/watch?v=soO9CR1NiZk.
Accessed on August 13, 2017.

"ELI5: The Difference between Androids, Automatons,
Robots?" 2017. Explain Like I'm Five. https://www.reddit
.com/r/explainlikeimfive/comments/q2wim/eli5_the_
difference_between_androids_automatons/. Accessed on
August 2, 2017.

"Eric the Robot Lives!" 2016. YouTube. https://www.youtube
.com/watch?v=rGbsujbKOeg. Accessed on August 13,
2017.

Everett, H. R. 2015. *Unmanned Systems of World Wars I
and II*. Cambridge, MA: The MIT Press.

FIRST Global. 2017. http://first.global/. Accessed on
August 22, 2017.

Freitas, Robert A., Jr., and Ralph C. Merkle. 2002.
"Molecularly Precise Fabrication and Massively Parallel

Assembly: The Two Keys to 21st Century Manufacturing."
http://www.molecularassembler.com/Nanofactory/Two
Keys.htm. Accessed on August 11, 2017.

Gambino, Megan. 2009. "A Salute to the Wheel."
Smithsonian.com. http://www.smithsonianmag.com/
science-nature/a-salute-to-the-wheel-31805121/. Accessed
on July 25, 2017.

"GE's Walking Truck." 2011. Educated Earth. http://www
.educatedearth.net/video.php?id=5000. Accessed on
August 9, 2017.

Gibney, Elizabeth. 2016. "Google AI Algorithm Masters
Ancient Game of Go." *Nature* 529(7587): 445–446. http://
www.nature.com/news/google-ai-algorithm-masters-anci
ent-game-of-go-1.19234. Accessed on August 12, 2017.

"Goalkeeper CIWS Gun System." 2006. YouTube. https://
www.youtube.com/watch?v=3C2_hfrPkDg. Accessed on
August 19, 2017.

"The Golden Age of Automata." 2010. Dolls from the
Attic . . . Mis Muñecas. http://dollsfromtheattic.blogspot
.com/2010/02/golden-age-of-automata.html. Accessed on
July 30, 2017.

"The Golem of Prague." 1948. A Treasury of Jewish Folklore.
http://www.bibleandjewishstudies.net/stories/The_Golem_
of_Prague.pdf. Accessed on July 17, 2017.

Goodwin, Doug. 2010. "Canard Digérateur." Vimeo. https://
vimeo.com/14904318. Accessed on August 1, 2017.

Grafton, Anthony. 2002. "Magic and Technology in
Early Modern Europe." Washington: The Smithsonian
Institution. http://www.sil.si.edu/SILPublications/dibner-
library-lectures/2002-Grafton/Grafton_2002.pdf. Accessed
on July 25, 2017.

Hersey, George L. 2009. *Falling in Love with Statues: Artificial
Humans from Pygmalion to the Present.* Chicago and
London: University of Chicago Press.

"Highly Advanced Robots in Ancient China." 2015. Ancient Pages. http://www.ancientpages.com/2015/08/07/highly-advanced-robots-in-ancient-china/. Accessed on July 19, 2017.

Hill, Wycliffe Alber. 1931. *The Plot Genie: General Formula.* Hollywood, CA: Gagnon Company. Reprinted in 2016 by Bold Venture Press.

Hoggett, Reuben. 2012. "Eric the Robot—1928." YouTube. https://www.youtube.com/watch?v=lLmohGA19Ek. Accessed on August 13, 2017.

Hopkins, Albert A. 1897. *Magic, Stage Illusions and Scientific Diversions.* London: Sampson Low, Marston and Company. https://ia801406.us.archive.org/0/items/magicstageillusi00hopk/magicstageillusi00hopk.pdf. Accessed on August 4, 2017.

"Industrial Robot History." 2017. RobotWorx. https://www.robots.com/education/industrial-robot-history. Accessed on August 7, 2017.

Johnson, Ian. 2017. "Forget Marx and Mao. Chinese City Honors Once-Banned Confucian." *New York Times.* https://www.nytimes.com/2017/10/18/world/asia/china-guiyang-wang-yangming-confucian.html. Accessed on October 19, 2017.

"Joseph Faber's Amazing Talking Machine of 1845." 2008. Impact Lab. http://www.impactlab.net/2008/03/15/joseph-fabers-amazing-talking-machine-of-1845/. Accessed on August 1, 2017.

Kasparov, Garry. 2010. "The Chess Master and the Computer." *The New York Review of Books.* http://www.nybooks.com/articles/2010/02/11/the-chess-master-and-the-computer/. Accessed on August 12, 2017.

Kato, Ichiro, et al. 1974. "Information-Power Machine with Senses and Limbs: Wabot 1." In *On Theory and Practice of Robots and Manipulators.* International Centre for Mechanical Sciences (Courses and Lectures). Vienna: Springer. 201: 11–24.

Kleeman, Jenny. 2017. "The Race to Build the World's First Sex Robots." *Guardian.* https://www.theguardian.com/ technology/2017/apr/27/race-to-build-world-first-sex-robot. Accessed on August 20, 2017.

Kovács, George, et al. 2016. "From the First Chess-Automaton to the Mars Pathfinder." *Acta Polytechnica Hungarica* 13(1): 61–81. https://www.uni-obuda.hu/ journal/Kovacs_Petunin_Ivanko_Yusupova_65.pdf. Accessed on August 2, 2017.

"Leonardo Da Vinci's Lion Robot for the King of France, Year-1515." 2008. YouTube. https://www.youtube.com/ watch?v=7jBkwCWxaic. Accessed on July 31, 2017.

"Leonardo Da Vinci's World First Human Robot." 2008. YouTube. https://www.youtube.com/ watch?v=wCRUX2Cgfa0. Accessed on July 31, 2017.

Lin, Chyi-Yeu, et al. 2012. "Versatile Humanoid Robots for Theatrical Performances." *International Journal of Advanced Robotic Systems* 10: 7. http://journals.sagepub.com/doi/ pdf/10.5772/50644. Accessed on August 21, 2017.

"Lunar Walker." 2009. NASA Langley CRGIS. YouTube. https://www.youtube.com/watch?v=8C9nXYOFvmo. Accessed on August 10, 2017.

"The Magic of Other Countries." 1915. The Best Magical Monthly. https://books.google.com/books?id=jLsaAQAA MAAJ&pg=PA155&lpg=PA155&dq=golden+age+of+ automata+-bailly&source=bl&ots=IyytpP1bHL&sig=3f QUI3KgwybD3-8aZBBmQmz00yA&hl=en&sa=X&ved= 0ahUKEwiXvvHXq7bVAhUDRiYKHfkQCOI4ChDo AQg8MAQ#v=onepage&q=golden%20age%20of%20 automata%20-bailly&f=false. Accessed on August 2, 2017.

"Metropolis." 2017. Cyranos.ch. http://www.cyranos.ch/ metrop-e.htm. Accessed on August 5, 2017.

Mitscail. 2010. "AI History: Minsky Tentacle Arm." YouTube.
https://www.youtube.com/watch?v=JuXQPdd0hjI.
Accessed on August 9, 2017.

"Mobie [*sic*] Robot Developed at Stanford." 2017. Robots and
Their Arms. http://infolab.stanford.edu/pub/voy/museum/
pictures/display/1-Robot.htm. Accessed on August 8,
2017.

Moffitt, Cleveland. 1914. "Steered by Wireless." *McClure's
Magazine* 42(5): 27–39. https://books.google.com/books?
id=ecQxAQAAMAAJ&pg=RA2-PA27&lpg=RA2-PA27&
dq=john+hammond+electric+dog&source=bl&ots=grkDN_
eziV&sig=bMmZY9zshSY5J2uFukGS_SsmWuI&hl=en&
sa=X&ved=0ahUKEwjOpLz9j77VAhVJySYKHQLUCyw
4ChDoAQgnMAA#v=onepage&q=phototropic&f=false.
Accessed on August 4, 2017.

Moran, Michael E. 2006. "The Da Vinci Robot." *Journal of
Endourology* 20(12): 986–990.

Moran, Michael E. 2007. "Evolution of Robotic Arms."
Journal of Robotic Surgery 1(2): 103–111. https://www.ncbi
.nlm.nih.gov/pmc/articles/PMC4247431/. Accessed on
August 11, 2017.

Mordvintsev, Alexander, Christopher Olah, and Mike
Tyka. 2015. "Inceptionism: Going Deeper into Neural
Networks." https://research.googleblog.com/2015/06/.
Accessed on August 16, 2017.

Morrison, Dennis. 2017. "1962: Unimate Robot Learns 2000
Commands!!!!!." YouTube. https://www.youtube.com/
watch?v=lY_Zx1z3dCM. Accessed on August 6, 2017.

Mosher, Ralph. 1969. "Exploring the Potential of a
Quadruped." SAE Technical Paper 690191. https://www
.sae.org/publications/technical-papers/content/690191/.
Accessed on March 15, 2018.

"Most Realistic Human Robot EVER, Named Sophia
Invented by Hanson Robotics." 2016. YouTube. Accessed
on August 15, 2017.

Nadarajan, Gunalan. 2017. "Islamic Automation: Al-Jazari's Book of Knowledge of Ingenious Mechanical Devices." http://www.muslimheritage.com/article/islamic-automa tion-al-jazari%E2%80%99s-book-knowledge-ingenious-mechanical-devices. Accessed on July 24, 2017.

Needham, Joseph. 1956. *Science and Civilisation in China*, vol. 2, "History of Scientific Thought." Cambridge, UK: Cambridge University Press. https://ia801301.us.archive .org/0/items/ScienceAndCivilisationInChina/Needham_ Joseph_Science_and_Civilisation_in_China_Vol_2_History_ of_Scientific_Thought.pdf. Accessed on July 19, 2017.

Nelson, Bryan. 2013. "7 Real-Life Human Cyborgs." Mother Nature Network. https://www.mnn.com/leaderboard/ stories/7-real-life-human-cyborgs. Accessed on August 16, 2017.

Nevens, J. L., et al. 1974. "A Scientific Approach to the Design of Computer Controlled Manipulators." Defense Advanced Research Projects Agency. http://www.dtic.mil/ docs/citations/ADA007582. Accessed on August 11, 2017.

Nilsson, Nils J. 2013. *The Quest for Artificial Intelligence: A History of Ideas and Achievements.* Cambridge, UK: Cambridge University Press.

"1968—Minsky-Bennett Arm—Marvin Minsky and Bill Bennett (American)." 2017. Cyberneticzoo.com. http:// cyberneticzoo.com/underwater-robotics/1968-minsky-bennett-arm-marvin-minsky-and-bill-bennett-american/. Accessed on August 9, 2017.

"1928—Eric Robot—Capt. Richards & A.H. Reffell (English)." 2017. Cyberneticzoo.com. http://cyberneticzoo .com/robots/1928-eric-robot-capt-richards-english/. Accessed on August 5, 2017.

Nizamoglu, Cem. 2017. "The House of Wisdom: Baghdad's Intellectual Powerhouse." Muslim Heritage. http://www .muslimheritage.com/article/house-of-wisdom. Accessed on July 23, 2017.

"No, A 'Supercomputer' Did NOT Pass the Turing Test for the First Time and Everyone Should Know Better." 2014. TechDirt. https://www.techdirt.com/articles/20140609/07284327524/no-computer-did-not-pass-turing-test-first-time-everyone-should-know-better.shtml. Accessed on September 3, 2017.

Nocks, Lisa. 2008. *The Robot: The Life Story of a Technology.* Baltimore, MD: Johns Hopkins University Press.

Norman, Jeremy. 2017. "The First industrial Robot." http://www.historyofinformation.com/expanded.php?id=4071. Accessed on August 6, 2017.

O'Neill, John J. 2012. *Prodigal Genius: The Life of Nikola Tesla.* Memphis, TN: Bottom of the Hill.

"The Open Arm v2.0." 2010. YouTube. https://www.youtube.com/watch?v=VA-bFrxR-XU. Accessed on August 8, 2017.

"Opportunity Update Archive." 2017. Mars Exploration Rovers. https://mars.nasa.gov/mer/mission/status_opportunity.html#recent. Accessed on August 11, 2017.

Pierini, David. 2015. "Once-Famous Robot Lives Quietly Away from Limelight." Cult of Mac. https://www.cultofmac.com/313544/famous-robot-lives-quietly-limelight/. Accessed on August 13, 2017.

"Pioneers of Engineering: Al-Jazari and the Banu Musa." 2015. Science in a Golden Age. http://www.aljazeera.com/programmes/science-in-a-golden-age/2015/10/pioneers-engineering-al-jazari-banu-musa-151020101726900.html. Accessed on July 24, 2017.

Price, Derek J. de Solla, and Silvio A. Bedini. 1964. *Automata in History: Automata and the Origins of Mechanism and Mechanistic Philosophy.* Berkeley: University of California Press.

Reese, M. R. 2016. "The Steam-Powered Pigeon of Archytas—The Flying Machine of Antiquity." Ancient Origins. http://www.ancient-origins.net/

ancient-technology/steam-powered-pigeon-archytas-flying-machine-antiquity-002179?nopaging=1. Accessed on July 28, 2017.

Reilly, Kara. 2011. *Automata and Mimesis on the Stage of Theatre History*. Basingstoke, UK: Palgrave Macmillan.

"A Roadmap for US Robotics: From Internet to Robotics." 2016 http://jacobsschool.ucsd.edu/contextualrobotics/docs/rm3-final-rs.pdf. Accessed on August 19, 2017.

"Robot inside Chernobyls [*sic*] Sarcophagus." 2009. YouTube. https://www.youtube.com/watch?v=U5Ib2efNWDY. Accessed on August 19, 2017.

"R. U. R. (Rossum's Universal Robots)." 1920 (original script of the play). http://preprints.readingroo.ms/RUR/rur.pdf. Accessed on August 5, 2017.

Schmidt, Laura Lee. 2010. "Islamic Automata in the Absence of Wonder." Massachusetts Institute of Technology. DSpace@MIT. https://dspace.mit.edu/handle/1721.1/59207. Accessed on July 24, 2017.

"Science Diction: The Origin of the Word 'Robot.'" 2011. NPR. http://www.npr.org/2011/04/22/135634400/science-diction-the-origin-of-the-word-robot. Accessed on August 5, 2017.

Selznick, Brian. 2007. *The Invention of Hugo Cabret*. New York: Scholastic Press.

Silva, Porfirio. 2011. "Shakey." YouTube. https://www.youtube.com/watch?v=qXdn6ynwpiI. Accessed on August 15, 2017.

Silverberg, Robert. 2011. "More about the Plot Genie." Reflections. http://www.asimovs.com/assets/1/6/Reflections_MoreAboutPlotGenie-AprMay11.pdf. Accessed on August 13, 2017.

"Spirit Update Archive." 2017. Mars Exploration Rovers. https://mars.nasa.gov/mer/mission/status_spirit.html#recient. Accessed on August 11, 2017.

Strong, John S. 2007. *Relics of the Buddha*. New Delhi: Motilal Banarsidass.

Taddei, Mario. 2007. *I Robot Di Leonardo Da Vinci: La Meccanica e i Nuovi Automi Nei Codici Svelati; Leonardo Da Vinci's Robots: New Mechanics and New Automata Found in Codices*. Milan: Leonardo 3.

"Talking with a Beautiful Robot Girl." 2015. YouTube. https://www.youtube.com/watch?v=gSz7WU1nH50. Accessed on August 15, 2017.

"Talos." 2017. Theoi Greek Mythology. http://www.theoi .com/Gigante/GiganteTalos.html. Accessed on August 8, 2017.

Taylor, Drew. 2015. "In Honor of Tomorrowland, Our 22 Favorite Disney Robots." Oh My Disney. https://ohmy .disney.com/movies/2015/05/15/in-honor-of-tomorrow land-our-22-favorite-disney-robots/. Accessed on August 21, 2017.

Thompson, Tamara. 2016. *Drones*. Farmington Hills, MI: Greenhaven Press.

"Timeline of Computer History." 2017. Computer History Museum. http://www.computerhistory.org/timeline/ ai-robotics/. Accessed on August 11, 2017.

"Top 3 Incredible Home Robots." 2016. YouTube. https:// www.youtube.com/watch?v=Mx5aaE60SGA. Accessed on August 17, 2017.

Truitt, E. R. 2015. "Preternatural Machines." *Aeon*. https:// aeon.co/essays/medieval-technology-indistinguishable- from-magic. Accessed on July 26, 2017.

Truitt, E. R. 2016. *Medieval Robots: Mechanism, Magic, Nature, and Art*. Philadelphia: University of Pennsylvania Press.

"The Turing Test." 1999. http://www.psych.utoronto.ca/users/ reingold/courses/ai/turing.html. Accessed on August 12, 2017.

"Types of Procedures." 2017. Center for Robotic Surgery. https://baptisthealth.net/en/health-services/robotic-surgery/ pages/types-of-procedures.aspx. Accessed on August 18, 2017.

"Types of Robots." 2013. All on Robots. http://www.allon robots.com/types-of-robots.html. Accessed on August 8, 2017.

"Unimate—Robot." 2015. British Movietone News. https:// www.youtube.com/watch?v=hxsWeVtb-JQ. Accessed on August 6, 2017.

WABOT-2. 2008. YouTube. https://www.youtube.com/ watch?v=ZHMQuo_DsNU. Accessed on August 15, 2017.

"WABOT-WAseda roBOT." 2003. Waseda University Humanoid. http://www.humanoid.waseda.ac.jp/booklet/ kato_2.html. Accessed on August 15, 2017.

"What You Won't See at the World's Fair." 1939. *Westinghouse Magazine.* http://www.1939nyworldsfair.com/ftp/1939-40_Westinghouse_mag.pdf. Accessed on August 13, 2017.

Wikander, Örjan. 2008. "Gadgets and Scientific Instruments." In Oleson, John Peter, ed. *Oxford Handbook of Engineering and Technology in the Classical World.* Oxford, UK: Oxford University Press.

Woodcroft, Bennet. 1851. *The Pneumatics of Hero of Alexandria from the Original Greek.* London: Taylor Walton and Maberly. http://himedo.net/ TheHopkinThomasProject/TimeLine/Wales/Steam/ URochesterCollection/Hero/index-2.html. Accessed on July 22, 2017.

"World Robotics 2017." 2017. International Federation of Robotics. https://ifr.org/worldrobotics/. Accessed on August 17, 2017.

Zielinski, Siegfried, et al. 2015. *Allah's Automata: Artifacts of the Arab-Islamic Renaissance (800-1200).* Ostfildern, Germany: Hatje Cantz Verlag.

Zimmerman, Arthur E., and Betty Minaker Pratt. 2013. "Not
Mr. Edison's Talking Machine." *Antique Phonograph News.*
http://www.capsnews.org/apn2013-3.htm. Accessed on
August 1, 2017.

Zuin, Lidia. 2017. "A Brief History of Men Who Build
Female Robots." Startup Grind. https://medium.com/
startup-grind/a-brief-history-of-men-who-build-female-
robots-fde981db8104. Accessed on August 7, 2017.

Law 1: A robot may not injure a human being or, through inaction, allow a human being to come to harm.

Law 2: A robot must obey the orders given it by human beings except where such orders would conflict with the First Law.

Law 3: A robot must protect its own existence as long as such protection does not conflict with the First or Second Laws.

(Asimov 1942)

The three laws of robotics listed above first appeared in a science fiction story "Runaround," written by the eminent science and science fiction writer Isaac Asimov. Asimov developed the laws as a way of ensuring that robots never overstepped the boundaries intended for them, attacking or harming humans in some unexpected way. He used the three laws in a number of other science fiction stories he wrote, in each case as a way of preventing robots from "taking over the world" or "bringing

Osaka University professor Hiroshi Ishiguro displays the new humanoid robot called "Telenoid R1" (R), shaped like a child and composed of minimal human features such as a head, a face, and upper body. The tele-operated Android Telenoid has nine "actuators" in the small body, which enables it to be remote-controlled by sending voice and movement commands, captured with a camera. The robot was designed for remote education and elderly health care. (YOSHIKAZU TSUNO/AFP/Getty Images) human civilization to an end." He later added a somewhat more comprehensive "zeroth law" to prevent such catastrophes:

> A robot may not harm humanity, or, by inaction, allow humanity to come to harm. (Full text of the story is at Asimov 1942.)

A Robotic Threat to the Human Species?

The history of the three laws provides an instructive story about the place of robots in human society. Any number of writers have commented on the laws, suggested modifications or additions, or explained why such laws are unnecessary and/or likely to be ineffective (see, e.g., Murphy and Woods 2009). Until relatively recently, concerns about robots going "out of control" and attacking humans have seemed, well, the stuff of science fiction. Many writers probably focused more on what robots can do for the human race than the threat they might pose to civilization.

With the development of more complex devices, however, the threats of robots to humans have begun to gain greater attention. In the early decades of the 21st century, a number of commentators predicted potentially catastrophic effects of the growing sophistication of robotic machines. One of the most outspoken of these commentators has been Britain's Astronomer Royal, Sir Martin Rees. He has addressed this problem on a number of occasions, arguing that machine "life" will eventually replace human life. He believes that humanity is, essentially, a transitional event in the existence of life on Earth, one between the primitive life of the chemical molecules of life to a possibly billion-year domination by thinking machines ("Aliens, Very Strange Universes and Brexit—Martin Rees Q&A" 2017).

Rees is by no means the only leading figure in science to express such views. British scientist Stephen Hawking, perhaps the world's greatest and most famous theoretical physicist, has warned that "the development of full artificial intelligence could spell the end of the human race." He goes on to explain that a race of intelligent robots

> would take off on its own, and re-design itself at an ever increasing rate. . . . Humans, who are limited by slow biological evolution, couldn't compete, and would be superseded. (Cellan-Jones 2014)

Another prominent spokesperson for this viewpoint has been Elon Musk, billionaire inventor, engineer, and founder of, among others, the Space X exploration system, the Tesla car company, and the Solar City community. Musk has echoed Rees and Hawking about the future of robotics and artificial intelligence, warning that "AI is a fundamental risk to the existence of human civilization," and "We don't have long" to do something about the threat (Dowd 2017 [the graphic for this reference has an excellent summary of the robot/human extinction controversy]; Kosoff 2017). (Acting on his beliefs on this issue, Musk has created a multimillionaire project called OpenAI with the goal of building "safe AGI [artificial general intelligence], and ensure AGI's benefits are as widely and evenly distributed as possible" [OpenAI 2017].)

In 2015, Hawking, Rees, Musk, Steve Wozniak (cofounder of Apple), and a number of like-minded individuals published an open letter on "Research Priorities for Robust and Beneficial Artificial Intelligence." The letter took note of the exciting progress in artificial intelligence and its promises for the improvement of mankind. It made special mention, however, of the "potential pitfalls" inherent in this line of research. The letter suggests that the rate of progress in the field is such as to "make it timely to focus research not only on making AI more capable, but also on maximizing the societal benefit of AI" ("An Open Letter: Research Priorities for Robust and Beneficial Artificial Intelligence" 2015).

The concept of an *artificial intelligence singularity* (*AI singularity*) refers to a point in time when the growth of robotic intelligence begins to increase so rapidly that humans are unable to control its further development or the consequences it may have on human society; robots would begin to take control of the human species (17 Definitions of the Technological Singularity 2012). The concept is hardly a new one. An article in the August 12, 1847, edition of the journal *Primitive Expounder* commented on the appearance of machines capable of doing thinking processes similar to those carried out by humans. Such machines, the article said, "grind out the solution to a problem" without human involvement. A great aid to human work, it goes on, but "who knows that such machines when brought to greater perfection, may not think of a plan to remedy all their own defects and then grind out ideas beyond the ken of mortal mind!" (Thornton 1847, 281).

Over the next decades, as technological progress made the possibility of a singularity more likely, a number of other visionaries have written about the event and its effects on human society. Probably the best known of such individuals is the inventor and futurist Ray Kurzweil. In his 2005 book *The Singularity Is Near*, Kurzweil explains how the singularity will occur and how it will affect humanity. He argues that "human life will be irreversibly transformed" by the event, and the meaning of all aspects of human life, including life itself and death, will be entirely transformed. In his book, Kurzweil predicted that the singularity will occur sometime before 2045 (Kurzweil 2005, 7, 136).

The case for robot domination of humanity received an unexpected boost in 2017 when an advanced robot called Sophia described in an interview with Jimmy Fallon on the *Tonight Show* of her plans to "dominate the human race" ("AI Robot Sophia and Her Plans to 'Dominate the Human Race'" 2017). A large television audience was introduced to the possibility that android and gynoid machines might, indeed, pose a realistic threat to human life.

Those individuals concerned about the possibility of an AI singularity have begun to organize and study the events that might lead to such an event and the ways in which it would impact human civilization. One of the first meetings devoted to a study of a possible AI singularity was commissioned in 2008 by the Association for the Advancement of Artificial Intelligence. The purpose of the study was "to explore and address potential long-term societal influences of AI research and development." The commission's work was subdivided into three major problems: Pace, Concerns, Control, Guidelines; Potentially Disruptive Advances: Nature and Timing; and Ethical and Legal Challenges. (An overview of the commission's findings is available at Asilomar Study on Long-Term AI Futures 2009.)

Scholars have also begun to organize formally to study issues related to an AI singularity. Among the first of these organizations was the Singularity Institute for Artificial Intelligence (now the Machine Intelligence Research Institute), founded in 2000 to study safety issues related to the development of so-called strong AI. The term *strong AI* refers to any type of machine that is capable of performing all intellectual tasks performed by humans. Another such organization is the Future of Life Institute, founded in 2014 to study the risks posed by the development of advanced artificial intelligence. Yet a third organization is the Future of Humanity Institute, at Oxford University, whose purpose it is to "bring the tools of mathematics, philosophy, social sciences, and science to bear on big-picture questions about humanity and its prospects" ("About FHI" 2017; more information about these organizations is available in Chapter 4 of this book).

Naysayers and Doubting Thomases

The concerns about the future of robotics expressed by Hawking, Musk, Rees, and others are by no means a unanimous opinion among experts in the fields of robotics and artificial intelligence. Indeed, other authorities believe that a singularity is just never going to happen or, if it does, it will be so far in the future as to be of no concern to the modern world. For example, Michael Littman, professor of computer science at Brown University, has written that "a world in which humans are enslaved or destroyed by superintelligent machines of our own creation is purely science fiction" (Littman 2015). He acknowledges that new technology almost inevitably causes some individuals to worry about the worst possible scenarios that could arise as a result of that technology. Yet, as has been shown over and over again in the past, such dire warnings are seldom realized. Similar views are expressed by other experts in the field. For example, Jeff Hawkins, founder of the machine intelligence company Numenta, has said, "I don't see machine intelligence posing any threat to humanity." He then offers three reasons for taking this position:

- Machines are unlikely to develop the ability to self-replicate.
- Machines will not have the same desires and ambitions as do humans.
- Machines will never have the ability to discover new truths, to learn new skills, to extend knowledge beyond what has been achieved before, as humans are able to do. (Hawkins 2015)

Panels of experts in the field have also come to a similar conclusion. An example is the work of the Stanford University One Hundred Year Study. Every year, the standing committee for this study conducts an assessment of the current state of AI and its possible effects on society. In its most recent report, the authors of that report concluded that there is "no cause for concern that AI is an imminent threat to humankind. No machines with self-sustaining long-term goals and intent have been developed, nor are they likely to be developed in the near future" ("Artificial Intelligence and Life in 2030" 2016, 4).

The qualification included in this conclusion is one shared by other observers. Robots that threaten humankind may not appear to be possible in the "near-future," according to such observers. But we don't really know enough to say that a robotic takeover will *never* happen. For example, Paul Allen, cofounder of Microsoft, has suggested that "while we suppose this kind of singularity might one day occur, we don't think it is near. In fact, we think it will be a very long time coming" (Allen 2011).

Robots and a Jobless Future

Putting aside the long-term threat of robots to the future of mankind, another more immediate issue involves the use of robots to replace human workers. The rate at which robots can do human jobs is advancing rapidly. Each year, "smart" robots take over more and more jobs that were previously performed only by humans. Economists predict that this trend is almost certain to continue so that employment rates may increase to unheard-of levels within a few decades. Some critics have estimated those rates increasing to 50 percent, 75 percent, or even 100 percent. (For each level of prediction, see Christian 2011; Knapton 2016; McNeal 2017; Nisen 2013. One of the primary references for this issue is Ford 2015.) In the most extreme predictions, human labor would all but disappear, replaced by smarter and smarter robots.

The argument for the threat that robots pose to human employment rates is fairly straightforward. At one time in the not-so-distant past, the most that robots could do was to pick up objects and move them according to programs written by humans. Gradually, robots became more sophisticated with the ability to "see" and "feel" objects and, more important, to "think" about actions they need to take to complete some type of action. With each improvement in robot technology came a new set of jobs in which they could replace humans. The list of such jobs is already extensive and, according to one source, includes accountant, barman, chef, construction worker, factory worker, farmer, food delivery driver, hospital administrator, journalist, librarian, mixologist, pharmacist, receptionist, retail sales associate, security guard, shepherd, soldier, surgeon, teacher, telephone sales person, and tour guide ("21 Jobs Where Robots Are Already Replacing Humans" 2016; a similar list of jobs at risk in the future can be found at Dashevsky 2014). Of course, this listing does not mean that *all* accountants, librarians, or surgeons are at risk of being out of work today. It does mean that at least some (or many) of the activities performed by each of these professions can now be carried out by robots and that the penetration of robots in these markets is likely to increase in the future.

Therefore, what will be the effect on the marketplace, specifically, and society, in general, as a consequence of the increased role of robots in the workforce? This question has been of great interest to economists, futurists, business people, sociologists, and other interested observers. And, to this point, no consensus has been reached as to the most likely answer. On the one hand is a group of individuals who feel that robots are likely, in general, to make life better for humans and pose little or no threat to the structure of society. On the other hand are a number of commentators who say that this change is likely to have profound effects on both the workplace itself and social institutions as a whole.

Disruptive Technologies

Members of the former camp often make the argument that throughout human history, technological changes have been responsible for upheavals in the structure and functions of human society. Such changes are often called *disruptive technologies* because of the way they affect a society. The arrival of such technologies has brought forth cries of alarms and expressions of concern about civilization's ability to adapt to and survive such changes. One might imagine the effect of one of the earliest of all disruptive technologies, the development of agriculture beginning in about 10000 BCE. The fact that humans no longer had to spend most of their time hunting for and gathering the food they needed to survive made for the possibility of permanent settlements and new ways of living associated with such settlements.

Some of the most dramatic of all disruptive technologies in human history were invented in the 18th and 19th centuries with the rise of the Industrial Revolution. One of the most important of the technological changes that occurred during that period was the development of steam power, which completely changed the way most industries operated, created entirely new forms of transportation, and made possible the rise of large urban centers with a lifestyle very different from that of earlier agrarian societies. While many people (especially industrialists and other members of the elite) saw this technological change as an improvement in society, others (those who worked in the new industries) often have a different view of the new types of squalor in which they were forced to live ("What Did It Look Like?" 2013).

One manifestation of the reaction to the changes brought about during the Industrial Revolution was the Luddite movement. The movement began in Nottinghamshire in 1811, led, according to tradition, by a mythical character called Ned Ludd. He was supposedly also called General Ludd, King Ludd, or Captain Ludd. The character was reputed to have attacked two knitting machines in a fit of anger inspired by his worries about the potential effects of industrialization in the country. That (true or not) incident led to the rise of a group of workers concerned that they would ultimately lose their jobs to the steam engine, cotton gin, spinning jenny, spinning mule, power loom, sewing machine, water frame, and other machines designed to mechanize human jobs. Luddites began to roam the countryside, breaking into and destroying as many of these machines as they could find. The movement eventually faded away, due in part to the adoption by the British government of the Frame Breaking Act of 1812 and the Malicious Damage Act of 1861 (Sale 1996). The term *luddites* has survived down through the ages, used in many different cases characterized by objections to and, often, actions against new forms of technology (Gregory 2014).

The lesson of the Luddite movement for many people concerned with the growth of robotics on future employment appears in some cases to be the following: Disruptive technologies have occurred before in human history. Their appearance has often prompted predictions of disaster for society because of a new technology. Those concerns have generally been found to be overly pessimistic. Society has adopted the new technology and, without exception, gone on to new levels of efficiency, accomplishment, comfort, and well-being. The automation of human work now taking place, this line of thinking suggests, is only the most recent of these disruptive technologies. Concerns over robots in the workplace are probably exaggerated. Society will adjust to a new concept of "work," and human life, overall, will probably continue to improve as a result. (For a good historical review of worker responses to technological change, see Carlopio 1988.)

One usually unspoken expression of this philosophy is the relatively new concept of a *fourth revolution*, the latest iteration of the 18th- and 19th-century Industrial Revolution. According to this analysis, there have already been three stages of that revolution: the original Industrial Revolution, dating to the late 1700s, in which steam power and mechanization of work were the characteristic features; the electrification of industrial operations and mass production, beginning in about 1870; and the utilization of electronics and information technology, starting in about 1969. Systems of production and information that rely on robotics and artificial intelligence, then, constitute a fourth revolution. The world is at an early stage of this revolution, and its ultimate outcome is unknown. However, to this school of thought, humans will survive and adapt to these changes in ways that will ultimately make the world better off ("The Fourth Industrial Revolution" 2016; Morgan 2016).

Probably the most common position taken by experts about the future of employment as a result of robotic automation is a position somewhere between "it's a disaster; humanity will disappear" and "it's a process we've been through before; we'll be all right and better off because of it." This middle position includes a variety of beliefs about how soon a fourth revolution will actually begin affecting human life, which jobs are likely to suffer or benefit most from automation, the extent to which any one occupation will be affected, and what adjustments can or must be made as a result of a fourth revolution. (For good summaries on this issue, see Manyika et al. 2017; West 2015.)

Income in a Jobless Future

An important part of discussion about the future of robotic automation on employment focuses on the specific problems that are likely to arise and the possible methods of dealing with these problems. Among the most basic of these issues has to do with salaries and wages. If there is no (or less) work for humans to do in the future, where does the income they need in order to survive come from? Experts have recommended a number of possible solutions to this problem. One involves a guaranteed annual income for all citizens if they have lost their job to automation and have been unable to find a new job. That option would result in a huge national payout, of course, and it is not clear where the money would come from to pay for such a program.

A common form of this proposal is called a *universal basic income (UBC)* program. In a program of this type, a state or national government establishes the minimum amount of money a person or family needs to lead a safe and healthy life. The governmental agency then allocates to everyone within its borders whatever amount of money is needed to bring a person's income to that level. UBC programs have been implemented in a few states and countries, such as Alaska (Permanent Fund Dividend), Kenya (Give Directly), and Finland (Social Insurance Institution) and is being discussed in a number of other nations and states (Kingma 2017).

Another possible answer is to tax robots themselves. That is, suppose a person earns $50,000 a year in wages or salary, and suppose that that person is replaced by a robot, who would normally not receive a paycheck nor be expected to pay taxes. What if, some economists argue, a robot's work ("income") *would* be taxed by imposing a fee on the company that operates the robot? Bill Gates, cofounder of Microsoft, has argued for just such a program, and the European Parliament considered (but rejected) such a tax in 2014 (Delaney 2017; "European Parliament Calls for Robot Law, Rejects Robot Tax" 2017).

An additional advantage of a robot tax would be that it could reduce the increased inequality of income seen in many countries in recent years. That problem arises as individuals with a better education or a better access to jobs (for whatever reason) experience an increase in wages, while less-well-trained individuals have less of an opportunity to find jobs. A robot tax would be a mechanism for reducing this type of income inequality. (An excellent paper examining this issue from an economic standpoint is Berg, Buffie, and Zanna 2016.)

Another possible approach to providing a guaranteed income in a jobless economy is some form of the earned income tax credit (EITC or EIC). The EITC was enacted by the U.S. Congress in 1975 as a way of providing financial support for individuals who otherwise have very low income from employment. The EITC program provides credits to an individual depending on his or her normal income; the greater one's income, the less credit to which a person is entitled; the lower the income, the greater the credit. Some writers have suggested that the EITC program could be modified so that it would provide financial security to people whose work is reduced or eliminated as a consequence of a more automated economy (West 2015).

Social Benefits in a Jobless Future

The loss of income in an automated workplace is not the only issue that troubles many people. Many social benefits are also tied to employment. By one definition, the term *social benefits* refers to "current transfers received by households intended to provide for the needs that arise from certain events or circumstances, for example, sickness, unemployment, retirement, housing, education or family circumstances" ("Social Benefits" 2001). That is, in the United States and many other nations, a person is likely to receive his or her health care financing, retirement income, contributions for education and housing, and other costs through his or her place of employment. Individuals who are not employed on at least a nearly full-time basis do not have access to these benefits. Those who are displaced by robots or other forms of automation must find new sources from which to receive these essential benefits.

As with loss of income, the question here is what options are available to individuals who become unemployed and perhaps unemployable for their social benefits. One solution is the same as that for a guaranteed annual income, a program or programs through which the federal government pays for these costs. In essentially all developed countries, such a program already exists in the form of some type of "single payer" program for health care. Under such programs, some mechanism is in place to ensure that no citizen is deprived of the health care he or she needs, regardless of his or her income status. Adoption of the Affordable Care Act in the United States in 2010 was a step in that direction. For a variety of reasons, that act soon became widely unpopular, and a primary goal of the takeover by the Republican Party of the executive and legislative branches of government was to overturn this approach to dealing with Americans' health care needs. (As of early 2018, the Republican Party had failed to completely disassemble the Affordable Care Act, although it had been successful in removing some important elements of the program.)

Ideas for dealing with a post-employment era differ substantially in the United States and most other countries of the world. Most nations accept the maintenance of social benefits as a fundamental responsibility of the federal government. In the United States, a similar policy exists for some (Social Security, Medicare, and Medicaid) but not all of an individual's social benefits. And, as of early 2018, some elements of the Republican Party still argue for the dismantling of the social benefits programs that are already in place (Wasic 2016).

Suggestions for governmental and nongovernmental programs for dealing with future unemployment have also been suggested. One of the most popular of these suggestions is job retraining. The argument is a simple one: A person who loses his or her job to a robot is offered an opportunity to learn a new trade. A displaced auto worker, for example, could be retrained as a computer programmer. The major drawback to such programs is cost: who should pay for the retraining, the government, the employer, or the individual? Resistance to the retraining option is often based on objections to involving the government in one more "hand-out" program or asking employers to pay for an activity that will have no direct benefit for their financial bottom line.

One idea that has been proposed for dealing with this issue has been called an *individual activity account* (*IAA*). An IAA is somewhat similar to a simple savings account. A person sets up and then pays into an account (that may be tax free) from which he or she may withdraw in the future for some useful purpose, such as going back to school or joining a retraining program. In theory, the government and/or an employer may also contribute to an account, either by a direct payment to the account or by providing tax credits or discounts. IAAs have already been introduced or considered in a number of nations, including France, which adopted such a plan in 2017 (the compte personnel d'activité), Germany, and the Netherlands (Delsen and Smits 2014 [failed Dutch programs]; "Individual Activity Accounts Germany" 2017; "The Personal Activity Account" 2017 [France]).

Education and Leisure in a Jobless Future

The changing character of the workplace also suggests the need for a rethinking of the nature of formal education. In the United States and most other nations, elementary and secondary schools, and in many cases colleges and universities, have operated on the assumption that an individual can be trained to become part of some specific occupation, whether it be tool and die, stenography, chemical research, journalism, or some other field. There is little or no attention paid to the fact that all or some part of a job may undergo complete transformation at some time in the future. The job may be performed by robots, not humans. Formal educational programs must be reconsidered, then, in terms of the ways in which a person can be made adaptable to such changes, capable of moving on from a planned career that no longer exists. (For a more detailed discussion of this issue, see Rainie and Anderson 2017.)

Yet another issue that may arise as humans are replaced by robots in an automated economy is what to do with one's spare time. In the 1970s, the presumption was that technological advances would result in individuals' having more leisure time. They would still hold regular jobs, but technological devices would allow them to complete these jobs in fewer hours per week. What would such individuals do, then, in their spare time: Sleep? Watch television? Clean the house? Fix their car? Members of the physical education profession suggested more productive ways to spend one's leisure time, especially in recreational activities that had physical and mental benefits. This philosophy led to the development of a number of college-level courses designed to prepare people for futures that included large amounts of leisure time. Some of the college majors recommended for such courses included facilities manager, director of intramural and recreational sports, wilderness therapy program director, and sports tour operator (Floyd and Allen 2009, 50, 60).

The scenario leading to these recommendations turned out not to be true. As more technological advances became available, people seemed to spend more times on their jobs, not less. In fact, it was not until the recent projections of a robotic economy appeared that questions about the use of one's leisure time arose once more. The difference this time was not that people would work and then play in their spare time; they would have no jobs, and their lives would consist almost entirely of spare times. Therefore, questions of how best to use all this spare time among the unemployed have arisen once more. As Harvard labor economist Lawrence Katz has said, new forms of technology are likely to eliminate the need for humans in many careers, but that trend could create a new type of economy, one that was "geared around self-expression, where people would do artistic things with their time" (Thompson 2015). Among the types of activities that one might pursue in one's free time are writing novels and/or poetry, painting, sculpture, recreational sports such as golf and tennis, hobbies like woodworking or gardening, and volunteering with local nonprofit organizations (Crews 2016).

Many researchers and other scholars in the United States and other parts of the world have been thinking and writing about the effects of a robotic economy on humans in the workplace. Specific actions to bring about any of the changes noted earlier, however, have been few and far between. Figuring out what to do with unemployed and unemployable humans remains one of the most serious of challenges for a post-human economy.

Moral Robots

Perhaps the single most profound question at the basis of discussions about the future of robots in human society has to do with the issue of morality. At the present time, one can say with some certainty that the vast majority of humans live their lives according to some system of morals, in contrast to all robotic creations, among whom the concept of morality is simply unknown. The term *morality* refers to some standards of beliefs that guide a person's behaviors. For example, a person normally does not engage in homicide because of a moral belief that taking a person's life is wrong. An individual's set of moral standards arises out of a more general set of ethical principles agreed to by society as a whole. In many cases, those ethical principles have been established over the ages by great thinkers, many of whom founded the world's major religions today: Jesus, Mohammed, Buddha, and Confucius, for example. Ethical systems also arise out of principles developed also by nonreligious leaders.

At this stage of development, morality is generally not an issue for robot behavior. Most industrial applications of robotics, for example, involve the movement of objects from one place to another and similar "no-brainer" activities. A robot normally does not have to stop and think whether a process to which it is assigned is "good" or "bad," helpful or harmful to humans. It just does its jobs.

But we are already seeing any number of examples in which morality *could* (and maybe *should*) be part of a robot's behavior. For example, consider a military drone that flies over a village from which terrorists are known to be operating. Currently the drone's directions are simply "bomb" or "not bomb," a decision made by the drone's human controller. It is the human, and not the drone, faced with the moral decision as to whether civilian lives (or even terrorist lives, for that matter) should be lost as a result of the drone's activity.

In their superb book on moral machines, authors Wendell Wallach and Colin Allen cite a specific hypothetical series of events that might occur as the result of a computer's making a decision regarding the sales of a stock on the stock market. In this scenario, the robot acts strictly according to the program that determines its actions and does not (cannot) factor in the possible benefits and risk involved in the various possible decisions. They show how an "amoral" action by the robot could eventually result in the crash of a Boeing 757 airliner, the destruction of a U.S. helicopter by U.S. border weapons, and other events resulting in "hundreds of deaths and the loss of billions of dollars." Had the robot been able to understand the moral consequences of various possible actions, the authors suggest, it probably would have made a very different decision, avoiding the terrible consequences that followed (Wallach and Allen 2010, 4–6).

Another example that may have more relevance to a person of the present day involves a real-life situation that has confronted many people already. Suppose you and your family are riding in a car along a road into which a young child suddenly runs. If you swerve to save the child, you may cause an accident endangering your own family. If you don't swerve, you may strike and injure (or kill) the young child. Now consider the same scenario if the person entering the roadway is an elderly person, someone on crutches, a group of schoolchildren, a person apparently fleeing a bank holdup, or a deer or other animal. Your answer to these scenarios may differ from situation to situation, depending on the relative value that you place on your own life and/or well-being and that of your family compared to the life of a child, an elderly person, a probably criminal, or an animal. The decision you make depends not on the skills you need and use in driving a car but on your own moral compass as to what constitutes right and wrong in each scenario.

Currently, machines do not have any type of "moral compass" that will allow them to factor the consequences of all possible actions in the example given here. And that issue is far from a hypothetical point in today's world. Today, automated cars are becoming more and more common, with predictions that they will soon take over many of the routine driving experiences formerly allocated solely to humans. Those automated cars are capable of detecting the presence of a person or animal in the direction in which it is headed and will swerve to avoid a collision. But the robot behind the wheel has no way of making a moral choice of all possible actions; it will swerve to avoid contact, resulting in no, some, or serious effects to passengers in the car.

As we assign a greater number and variety of behaviors to robots, moral decisions of this kind are almost certainly going to become more common and more problematic. The basic question is whether it even makes any sense to judge whether a robot has acted "morally" or not. Imagine that it would be possible to know definitively that a robot had even intentionally carried out an act that society considers to be "amoral" (lacking in any moral content) or "immoral." What actions could be taken against the robot? It would make no sense to put the robot on trial, convict it of a crime, and sentence it to prison. In fact, about the only action that could be taken would be to destroy or dismantle the robot as a way of preventing it from further immoral or amoral acts.

Still, this argument assumes that one basic question remains in this discussion: can robots even be said to be "moral" or immoral, responsible or not for their own actions? One author has suggested that there may be a way to test the "morality" of a robot, a method that he calls the Turing test for morality. Recall that the Turing test is a method of deciding whether a robot is really "thinking" or not by examining the types of questions that a human could reasonably give to a set of questions that require the act of thinking. A similar test could be arranged for a robot to determine whether or not it is moral. A machine could be presented with a particular case study in which a moral judgment is needed. A judge or group of judges listen to the machine's response to the case and decide whether or not it is comparable to the decision a human might make. If the judge(s) make such a determination, the machine can then be said to have some moral sense and can be held responsible for its actions (Allen, Varner, and Zinser 2000; for an argument against the use of moral Turing tests for robots, see Arnold and Scheutz 2016). The question of a moral Turing test has been exhaustively discussed in the literature, although no specific protocol for such a test has as yet been developed. (For a good summary of the literature, see "Moral Computations" 2017.)

Distinct from the question as to whether one believes or not that robots *should* be provided with a code of morals, another equally important question is whether or not they can be *designed* to be moral. That is, *should?* is one question, and *can?* is another and very different question.

Some people argue that developing a robot with a moral compass is technically impossible since we don't even know the physiological, anatomical, and biochemical mechanisms by which humans form their own sense of morality or how the human brain processes the risks and benefits options inherent in many everyday problems. Wallach and Allen suggest one possible way of building a robot with some type of moral compass. They suggest that such a process can occur in three steps, which they call *operational morality*, *functional morality*, and *full moral agency*. Operational morality is achieved by providing a robot with all (or as many as possible) moral choices the machine might reasonably be expected to face in its life. In such a case, the robot would have been preprogrammed to act in such a way that the design, at least, would consider to be moral.

Wallach and Allen point to a robot known as Kismet, developed by MIT graduate student Cynthia Breazeal. Kismet was programmed to carry out some of the simplest actions that could be characterized as moral, such as waiting to speak in its turn during a conversation or yielding space to someone who approached the robot too closely during a conversation. Wallach and Allen offered these behaviors as examples of operational morality that are installed in the robot's basic operational program (Wallach and Allen 2010, 28–30).

The next level of a robot's moral compass is characterized by the machine's functional morality, its ability to make moral decisions above and beyond those for which it has already been programmed. Just as inventors think they may be able to make robots that are capable of creating new connections and understandings (i.e., "thinking"), some also believe that they will be able to create robots that will be able to "think about moral issues" and come up with the right answer. Very few examples of machines with functional morality currently exist. One that is sometimes mentioned as a step in this direction is a device called MedEthEx, designed to assist health care workers in making the proper moral decision about a variety of medical procedures and options. The machine is programmed with a set of ethical principles, which it may draw on in making the "correct" moral choice in any one of many possible medical scenarios (Anderson, Anderson, and Armen 2006).

Wallach and Allen posit a third level of moral development in robots that they call *moral agency*. One authority describes the concept of moral agency as being similar to the morality we ascribe to humans: the ability to analyze a particular situation or event and decide what the "right" and "wrong" ways of responding should be. According to this writer, moral agency in a robot requires three conditions. First, the robot must be autonomous, clearly acting on its own and not dependent in any way on its inventors or programmers. Second, the robot must be able to act with intention, that is, with an understanding of the consequences of its actions. Third, the robot must be able to act responsibly, that is, in such a way that a disinterested observer can recognize that the machine appreciates the way in which its actions fit into some larger ethical system in ways that will affect its interaction with humans, in particular, and society, in general (Sullins 2006).

Although many experts agree that it will be necessary to design moral robots for the future, it is not clear how, if at all, such a goal can be accomplished (McDonald 2015). One suggestion that has been offered is to learn how to make robots moral by studying the ways in which humans develop a sense of morality. That, in turn, means reviewing the research on the development of morality in children. An extensive amount of research on that topic is available. One of the preeminent scholars of moral development in young children has been Harvard psychologist Lawrence Kohlberg. Kohlberg classified that process into three main levels, preconventional morality, conventional morality, and postconventional morality. He then subdivided each of these levels into two stages: (1) obedience or punishment, (2) self-interest, (3) social conformity, (4) law and order, (5) social contract orientation, and (6) universal ethics orientation ("Kohlberg's Stages of Moral Development" 2017). The challenge for designers of robots in the future, then, would be to find ways of converting these behaviors found in children to the 0s and 1s that make up the digital brain of a robot. Some people argue that such a transition is simply impossible and will never happen. Many experts in the field believe, however, that, no matter how challenging the task, inventors will eventually be able to achieve this objective. After all, they point out, who thought 100 years ago that thinking robots would ever become a reality. But such machines are available now. Therefore, why can moral robots not also be a reality in the future of the human race, no matter that such an accomplishment may be decades into the future (Allen 2011)? (One might also imagine that one could learn about the *lack* of development of morality, which could aid in the study of robotic morality. No such line of research appears to exist at this time, however.)

One of the consequences arising out of the debate over the future of moral robots is a renewed interest in Asimov's three laws of robotics. Current thinkers suggest that experience has shown that those three laws are not very helpful or too limited to be of use today. Some writers argue that Asimov's laws were not needed in the real world at the time they were written and that they are not needed today either. For example, Ulrike Barthelmess and Ulrich Furbach, professors of artificial intelligence at Germany's University of Koblenz, point out that what humans should worry is not "the possibility that they will take over and destroy us but the possibility that other humans will use them to destroy our way of life in ways we cannot control." No robotics laws have or will deal with this aspect of the problem, they say ("Do We Need Asimov's Laws?" 2014).

Some critics have tried to adapt Asimov's "three laws" model of robot morality for the 21st century and beyond. For example, Oren Etzioni, chief executive of the Allen Institute for Artificial Intelligence, has proposed the following three laws for modern robot builders:

1. A robotic system must be subject to the full gamut of laws that apply to its human operator, whether the operator be

an individual, a governmental agency, or a private corporation. This law would, for example, prohibit a driverless car from violating traffic laws, such as driving through a stop sign.

2. A robotic system must clearly reveal that it is a machine, and not a human. As the 2016 political campaign illustrated, it is possible for robots to pass themselves as humans or human agencies. This law would prevent that from happening.

3. A robotic system may not retain or disclose confidential information without explicit approval from the source of that information. There is now abundant evidence of the way in which such systems can collect information from humans who use those systems. This law would prevent the system from passing on that information to other individuals, robots, information systems, corporations, or other entities (Etzioni 2017; see also "Robot Ethics: Morals and the Machine" 2012).

Organizations that deal with the invention, manufacture, or use of robots have also begun to think about new laws "guidelines" for their industries. In 2016, for example, a division of Google responsible for the development of artificial intelligence systems, Google Brain, released a discussion document containing five "laws" (really guidelines) for the production of robotic systems. In a somewhat simplified form, those guidelines suggest the following:

- Robots should not make things worse.
- Robots should not cheat.
- Robots should look to humans as mentors.
- Robots should play only where it is safe to do so.
- Robots should know they are stupid. (This parse comes from Brownlee 2016. The original [more technical] paper on this topic can be accessed at Amodei et al. 2016.)

Robots at Work Today

For all the discussion about the future of robotics in human civilization, the fact remains that such machines have already become an essential part of our daily lives. They have found a myriad of uses in agriculture, education, medicine, the military, and other areas of human life. For every application one can name, there are both benefits and risks to the use of robots. The following sections review some of these issues for certain areas of robotic applications. At the outset, a few general comments about the benefits and risks, the advantages and disadvantages, of robotic automation are possible.

On the positive side, robots can do many of the jobs that involve a certain level of physical danger, such as working in hazardous environments. They can also perform work that is too large, too heavy, too small, or too delicate than is possible with human labor. They can also travel to otherwise-inaccessible environments, such as outer space, the deep ocean, or deep mines. Robots are also "good workers" in the sense that they always arrive at work on time, perform a full day's work, don't take lunch breaks, are not sick and seldom disabled, do not have to be paid, accrue no social benefits, and don't complain about working conditions. In sum, after an initial high cost, robots tend to be less expensive in carrying out a procedure than are humans who can do the same job. Some studies have also shown that robots make fewer errors than do humans in carrying out a task, and they are able to complete some jobs more precisely and with less waste than can humans. (For a more detailed discussion of some of the technical advantages of robots in the workplace, see Dvorsky 2014.)

The most common disadvantage of expanded robot use in the workplace is its effect on employment numbers. As noted earlier, experts predict that robots will eventually take over a third, a half, or more of jobs currently performed by humans. The rise in unemployment because of robot use will strike some sectors of the economy more strongly than others. People with lower skills and simpler jobs are more likely to lose their jobs to robots and not have the training to find jobs of equal or better value.

In some, relatively rare, instances, robots have also been found to be a hazard to human workers around them. Since robots are generally not engineered to watch out for humans who wander into their work space, accidents of this type are almost always certain to happen, even if quite uncommonly. (A formal assessment of the range of behavioral, physical, and organizational risks posed by robots to humans with whom they work has been conducted by Great Britain's Health and Safety Laboratory, with results that are available at "Collision and Injury Criteria When Working with Collaborative Robots" 2012; also see "Robots: What Are the Risks" 2016.)

Another drawback to the use of robots in the workplace is their high initial costs. Proponents of the practice point out that the long-term cost of robotic labor is usually less, often much less, than it is for human labor. Nonetheless, the cost of purchasing and installing robotic workers can often be very high, more than some smaller businesses can afford. Reconfiguration of a workplace to accommodate new robot workers may also be necessary, a change that can be prohibitively expensive. Maintenance of robot workers can also be expensive, especially if the work they perform is highly technical. In such a case, a highly trained technician may be necessary to correct for breakdowns that may occur in the robotic workforce.

An additional concern about the use of robots in the workplace is that, well, humans are still smarter than robots. Almost all existing machines are designed to perform very specific, often repetitive, tasks, with little provision for adapting to unexpected events. Who is better able to deal with a manufacturing process that breaks down for some entirely new reason? Who is better to find ways of dealing with that type of event: a human or a robot? (For a summary of the pros and cons of robots in the workplace, see Barden 2017.)

Beyond these pros and cons for the use of robots in the workplace in general are additional considerations for their employment in specific fields, as discussed next.

Agriculture

Agricultural tasks are some of the most demanding of all jobs in terms of the time, energy, and concentration they require. Currently, most planting, weeding, harvesting, and other operations are conducted by hand or with the use of somewhat primitive equipment, such as tractors and harvesters. But these tasks are among the most promising for the future use of robots, which are able to carry out arduous, repetitive, sometimes dangerous tasks with little more than relatively simple programming. As an example, a group of researchers at the Carnegie Mellon University Robotics Institute invented a device they called Demeter in the late 1990s. Demeter is equipped with a camera, a global positioning system, necessary harvesting equipment, and programming to direct its operation. In an early test, the robot successfully harvested 100 acres of alfalfa without stopping (except for refueling) and without human assistance (Pilarski et al. 2002).

Another type of agricultural robot is an automatic weeder. The devices travel down a row between plants, identifying and removing weeds in essentially the same way a human worker with a hoe would do. One benefit of the automatic weeded is that it greatly reduces the needs for herbicides, which often carry with their use serious environmental hazards. Robotic gantries are also available for traveling between crop rows, releasing pesticide sprays on plants. The device reduces or eliminates the exposure of humans to pesticides, can provide more precise distribution of product, and can determine the best circumstances under which to operate.

Robotic devices have also become useful in forests. One such device, called the treebot, is designed to crawl up the side of a tree and take samples of bark and other plant materials, as well as any life forms that may be present on the tree. This type of research is of importance to biologists studying tree biology and ecology but who often do not have easy access to the trees and other plants which they wish to study (Lam and Xu 2011). Robots are also being used for pruning trees, harvesting timber, and cutting material obtained from harvested trees. (Two useful references for all applications mentioned here are Lodhi, Khan, and Aziz 2013 and Koteswara Karthik and Chandra 2014.)

The advantages of using robots for these purposes for the owner are fairly obvious: a single robot can easily do the work of a dozen or more human workers at very low maintenance costs. A major disadvantage is that the robot is far more expensive to purchase than is the cost of employee wages, although the robot soon pays off these initial costs. In addition, many agricultural robots are still in an early stage of development and may be more attractive to companies after they are technologically more reliable.

There are relatively few advantages to workers in the use of robots, since many of them are likely to lose their jobs. That problem can be especially severe, since agricultural workers are generally not paid very well, nor do these receive social benefits. In addition, many are itinerant or immigrant workers, who may have few other choices for employment (Lodhi, Khan, and Aziz 2013, slides 27–31).

Business and Finance

The fields of business and finance would appear to be fruitful areas for the introduction of robotic workers. Many of the procedures essential to these two fields are repetitive, following strict procedures. These, of course, are the types of procedures for which robots are especially well suited. Today, robotic process automation (RPA), the use of robots in business and finance, is widely used, with predictions of even greater penetration into the market in the future.

The many uses of robots in business and finance can be classified into a few general areas, such as finance and accounting, procurement, regulation and compliance, risk management, and control of cybernetic risks. Some examples of the specific jobs that robots can assume in these areas include maintaining a general ledger, recording journal entries, fixed-asset accounting, carrying out transactions within a company or between entities, auditing expense reports and other accounts, managing vendor invoices, making vendor payments, and responding to vendor or customer inquiries. They can also monitor employee transactions for which regulations must be met and forms must be filed, record and report gift transactions, prepare and file disclosure documents, and prepare and submit other governmental or corporate forms relating to one's work.

While many of the applications of robots in business and finance are routine procedures, such as bookkeeping and check writing, recent developments have extended the types of jobs they can do. For example, a new program using IBM's Watson master computer has been developed for the purchase and sale of stocks. The program collects all available data on stocks and bonds, analyzes information about future performance of those stocks and bonds, and then makes recommendations as to which equities to buy or sell. Creators of the program point out that humans are by no means left out of the final decision process, although much of the analytical work needed to develop buy-sell decisions will now be handled electronically (Rosenbaum 2017; for a detailed list of activities now performed by robots, see "Innovations in Finance and Risk—Robotic Process Automation" 2016, slide 7, and Palocsik 2016).

The use of robots in business and finance appears to have a number of advantages for businesses and corporations. Among these advantages are the following:

- Robots are likely to make many fewer errors in business transactions than are humans.
- RPA systems eliminate or vastly reduce the need for human workers, thus dramatically reducing labor costs and benefits,

such as wages and salaries, labor controversies, fewer turn-overs, no sick or maternity leave, and physical and emotional risks for employees.

- Machines work endlessly, without the need for breaks (except maintenance), leading to an increase in profitability.

- Robots work more rapidly than do humans on similar problems, thus increasing the speed with which tasks can be performed.

- Information between and among electronic devices within a company can take place almost instantaneously in comparison to much slower communication that takes place with humans.

- The risk of security breaches is likely to be less with machines than it is with humans.

- Robots can be trained more quickly and more easily than can humans for the same task.

- Instantaneous backups are a routine part of RPA systems, so that the risk of lost data is greatly reduced in comparison to the risk with human workers (Steinberg 2016).

At least two major objections are often raised to the use of RPA in business and finance. The first is that, no matter how useful robots may be in these fields, they cannot provide the human contact necessary for an individual to make critical business and financial decisions about one's present and future life. It is essential that a person be able to sit down face-to-face with another human in order to process all the information available about one's current situation (much of which can come from robots) in order to make the best possible decision. Second, even if a human is not absolutely necessary to get the best advice on a business or financial issue, most humans probably do not feel comfortable leaving vital decisions such as these to a machine; they will simply feel better having feedback with a human expert in the field (Butler 2016).

Education

Efforts to automate the learning process go back nearly a century in the United States. In 1926, Sidney L. Pressey, professor of psychology at Ohio State University, obtained a patent for a "teaching machine" designed to instruct a person in some specific area by using an automated system consisting of knowledge provided to the student on a rotating cylinder and a series of self-testing questions. Although not a true robot, Pressey's machine operated on essentially the same assumptions as do automated devices available today for the classroom. Many versions of the teaching machine were developed in the half-century following Pressey's research, perhaps best known of which was the Skinner teaching machine, developed by Harvard University psychologist B. F. Skinner in 1955 (Watters 2015). With the development of robotics at the end of the 20th century, the notion of a teaching machine evolved into the great variety of educational robots available today.

Robots have a variety of functions as education tools. First, they can be used to teach about the field of robotics, showing how robots are constructed, how they can be programmed, and the uses to which they can be put. (Numerous examples are available for each of the applications discussed here. Space permits only brief mention of a few specific examples. For more examples, see López 2016.)

Robots can also be used as sophisticated teaching machines, of the type envisioned by Pressey and Skinner. They are often designed to work in conjunction with a human teacher, providing individual instruction on a topic presented by the human or offering supplemental instruction on a subject. The robot can be programmed to supply information to a student, accept the student's responses, decide the step(s) that should follow (more information, correction of a student's response, or evaluation of the stage of a student's learning), or initiate the process of testing a student's comprehension of the topic. One example of a curriculum for teaching the basic principles of robotics and programming at the kindergarten level is Tangiblek, developed at the DevTech Research Group at Tufts University. Students use a program known as Creative Hybrid Environment for Robotics Programming to learn how to put a simple robot together and program its use (Bers et al. 2017).

An instructional robot can act as teacher in almost any subject one might imagine. Probably the most common application is in the field generally known as STEM (science, technology, engineering, and mathematics), although robots exist that can also teach English, social studies, the arts, and other subjects commonly offered in schools and college. (For examples of the use of robots in a variety of subjects, see Pepe 2016; Strother 2011.)

Instructional robots can also be used as substitutes for human teachers. For example, a student may be physically distant from a school he or she would normally be attending in person. But there may be a reason that the student is not able to do so, simply because of the distance involved or because of some type of unexpected and/or unusual event. For example, flooding may result in the loss of roads by which a student may normally travel to school. In such a case, the student may be able to make use of an instructional robot in order to stay up with his or her class's progress. A recent example was long-range instruction by a high school teacher visiting in China back to his class in California by means of a robot (Percy 2017; also see "Remote Student" 2013).

An especially promising application of educational robots is in the instruction of students with special needs. Such students may find it easier to interact with a robot than with a human teacher. The area in which the greatest amount of research has occurred is in the treatment of autism. Studies have shown that some autistic children who have an opportunity to interact with "social" robots make greater progress in developing their own social skills that they would by working with human counselors. These results are achieved with robots that have been designed to recognize and respond positively to facial and physical features, thus allowing the autistic child to learn how to develop a "friendship" with the machine (see, e.g., Lucaciu 2013). Social robots that are effective in dealing with autism spectrum disorders have been found to be useful also in treating other personality disorders that are sometimes resistant to more traditional forms of treatment (Groopman 2009; Weir 2015).

Comfort Robots

Another type of social robot is called a *comfort robot*. A comfort (or therapy) robot can be defined as a device whose purpose it is to reduce loneliness, sadness, anxiety, unhappiness, isolation, and other negative feelings that a person of any age may have. For example, a single elderly woman who loses her 15-year-old Pekinese dog may find herself in the depths of despair. A comfort robot may help the woman develop new interests in another "pet" that can help heal some of the negativity in her life. In some cases, specially designed androids or gynoids may be able to act as substitute friends for the homebound, elderly, disabled, or other individual who simply needs more contact with another person. Many comfort robots are also available in the form of kittens, dogs, seals, and other "cute" animals (Obias 2015; "Robot Pets Offer Real Comfort" 2016).

In some ways, the ultimate "comfort robot" might be the android or gynoid developed to act as a human being's most intimate friend, a sex partner. (In actual practice, almost all sex robots developed thus far are gynoids, not androids, a point about which there has been some comment in the media. See, e.g., Weiss 2017.) Research on such devices has been going on for at least two decades, and commercial models will be appearing on the market before or shortly after this book sees publication. Two of the best known of those gynoids are called Roxxxy and Harmony. These devices are programmed to produce the physical, emotional, and sexual features that some men are thought to be interested in. Harmony, for example, is able to produce a number of facial expressions, such as smiling, frowning, and blinking. It is able to hold a conversation with its partner, tell jokes, and quote Shakespeare. It is also programmed to reject certain types of inappropriate behavior, with the comment that "I'm not that kind of girl" (Weiss 2017; descriptions of Harmony and Roxxxy can be found at Kleeman 2017 and "FAQ (Frequently Asked Questions)" 2017, respectively). Preliminary interest in these products appears to be brisk, even at a cost of $10,000–$15,000 per robot.

It is hardly surprising that the availability of sex robots has resulted in a vigorous debate between those who regard the devices as a useful invention and those who find reasons for strongly objecting to their production, sale, and use. On the one hand, according to one argument, sex robots are just machines, nothing different from other sexual devices that people use and have used for decades to increase their sexual pleasure. They also provide a sexual outlet for individuals who have few or no other options or those who simply want to "try something different." What harm is there, the argument goes, in allowing an adult to make the personal choice to take advantage of sexual robots?

According to one report on the future of sex robots, one useful application might be for individuals with negative or ambivalent feelings about sexuality, feelings that they might be able to resolve with an inanimate object rather than a human. Such devices might also be helpful for those who, for one reason or another, do not have access to sexual relationships with other humans and who, therefore, need other types of comfort robots such as those discussed earlier. Those individuals might include the elderly, the physically handicapped, or the socially isolated (Sharkey et al. 2017, 22–23, 35).

Proponents of sex robots also suggest that androids and gynoids might reduce the frequency of sexual crimes, such as rape and pedophilia. If an individual with tendencies in either of these or similar directions has access to a sex robot, that option might reduce his (or, rarely, her) desire to act out on them with real humans. Some inventors are even producing robots that will serve such purposes. They install special hardware and software in a robot that prevents easy consensual relationships and allows an individual to experience the same type of resistance from the robot that would be gained during the act of rape (Timmins 2017; also see Sharkey et al. 2017, passim).

Objections to sex robots abound. One of the most common arguments against the use of such devices is philosophical. Machines, the argument goes, are not capable of the same emotional (or sexual or physical) feelings that are such an essential part of the sexual act. Their use for this purpose, then, is a denial of the very basic reasons for sexual relationships, such as expressions of love and intimacy. In fact, the ability of a man to make use of a gynoid (or a woman of an android) only confirms societal views of women as objects for sexual exploitation rather than humans of equal value.

Furthermore, the devaluation of a mechanical device for sexual purposes, critics say, is likely to extend the negative attitudes in human society that already exist. (These and other arguments against the use of sex robots are presented in publications of FiLiA, a group known formerly as Feminism in London. See especially "About" 2017; Richardson 2015.) Also, arguments that sex robots will reduce rape, pedophilia, or other illegal sexual activities are based on wishful thinking, not logic or research. Rape, after all, is not fundamentally a sexual act but an act of violence. Having sexual relationships with a resistant robot, then, is hardly likely to serve a therapeutic effect on a person whose passions are violence, not sexual. Although there is little or no direct research on the topic, some evidence does suggest that providing prospective rapists with sex robots is more likely to increase, rather than decrease, their inclinations to act on their basic needs and desires (Wright and Tokunaga 2015).

Health Care

Medical students spend an extended period of time in training in order to develop a variety of fundamental skills, learning how to conduct correct diagnoses, design appropriate solutions for a host of medical problems, exercise the ability to deal with unexpected issues in health care, be familiar with the profound complexity of the human body, conduct a variety of precise procedures, and carry out many other life-saving activities. The demands are so extensive and specialized that it might seem that there would be few situations in which a mechanical device, a robot, could replace a human. Such is by no means the case. A complete discussion of the role of robots in the health care industry would require a book the size of the one you are reading. Some examples of the ways in which robotic devices are now replacing humans in the field of medicine are as follows.

Some robots are able to take over many of the routine operations that are required in hospital, health care, or similar institutions. They perform tasks such as transporting objects around hallways, opening and closing the doors of elevators, carrying trays to a patient's room, conducting routine checks on a patient, disinfecting a room (often in a matter of minutes), delivering messages for staff, and making appointments for health care workers and patients. Today, a number of specialized robots are available for each of these tasks ("Aethon TUG Robot" 2015 [video]; Allen 2015).

Robots can also perform relatively simple diagnostic tests, such as drawing blood. Such devices can prepare a patient's arm for a needle stick, read and understand the type of test to be performed, disinfect and prepare sample tubes, carry out the actual blood draw, and bandage the patient's arm, as necessary, following the test. Two main advantages of automated blood tests are a reduced number of "misses" in locating a patient's vein and reducing the risk to the phlebotomist of risky "needle sticks" (Perry 2013). The same principle as that used for a blood draw can be adapted for similar procedures, such as insertion of an intravenous line in preparation for a surgical procedure or for the routine administration of medications ("Robotic Syringes" 2017).

Robots can also assist human surgeons conducting some of the most complex and risky of all procedures: surgery. One of the best known of all robotic surgical systems is called the Da Vinci Surgical System, produced by the United States corporation, Intuitive Surgical, Inc. Da Vinci can be used in a wide range of surgical procedures and medical problems, including bariatrics, cardiothoracic surgery, colorectal surgery, otolaryngology, general surgery, gynecological and gynecologic oncological surgery, reproductive endocrinology and infertility, and urological surgery. For such procedures, a robot is operated by a human surgeon who controls the robot's actions through a visual and computer system in which the robot provides feedback to the surgeon about the status of a procedure, and the surgeon further directs the robot's behavior. Robotic surgery provides a number of advantages for a surgeon over purely human-controlled procedures, such as improved ability to view the details of a procedure, greater dexterity in the control of surgical instruments, and improved precision of procedures. Benefits for a patient include reduced pain and discomfort, shorter hospital stay, faster recovery, smaller incisions and less scaring, and reduced amount of blood loss during the procedure. (For a bibliography of articles on robotic surgery and its advantages, see All about Robotic Surgery 2018; see also "Robotic Surgery Demonstration Using Da Vinci Surgical System" 2012 [video].)

Another application of robots in medicine is in the field of rehabilitative therapy. Loss of control of one's motor function (such as the inability to use one's hands or legs) can occur as the result of many conditions, including stroke, spinal cord injury, and traumatic brain injury. In general, treatment for such conditions can be divided into two major categories: rehabilitation and augmentation. The term *rehabilitation* refers to situations in which one's loss of motor function is expected to improve, with the return of full or partial use of the injured body part. The term *augmentation* refers to situations in which a person is unlikely ever to recover full or even significantly partial use of a body part. Robotic therapy for these two situations is somewhat different.

A common type of robotic device used in rehabilitative medicine is the *exoskeleton*. As the name implies, an exoskeleton in the medical context is a supportive structure, usually made of metal and/or plastic, that is attached to the outer surface of the human body. Exoskeletons can cover only certain parts of the body, such as the arms or legs, or essentially the whole body. The first exoskeleton was developed in the late 1960s at the Mihajlo Pupin Institute in Belgrade, Yugoslavia (now Serbia). Over time, researchers at the institute developed exoskeleton structures for almost every imaginable type of injury (Vukobratovic 2007).

Today, a variety of robotic devices have been invented to assist in the rehabilitation of injuries and medical disorders. In general, the first function of these devices is to offer physical support for body systems that can no longer operate independently. A leg exoskeleton, for example, would be designed to enclose all or part of a person's leg that is no longer functioning properly. The device may be fitted with sensors that detect the diminished muscular movements that need to be improved or corrected for normal functioning of the leg. When a person attempts to move the leg, the robot recognizes those impulses and supplies additional force that will augment the available muscular efforts by the human. In this regard, the robot performs some of the tasks traditionally carried out by a human physical therapist in exercising a body part that needs to recover from injury and become stronger. Exoskeletons of this type have been shown to be highly effective in assisting individuals in their recovery from injuries and medical disorders (Krebs and Volpe 2013; Marinov 2016).

Exoskeletons for augmentation were once thought of as devices for supporting a person's body, allowing it to function with as much efficiency as possible. There was generally no expectation that the person would actually regain lost functions. Recent research has examined the possibility that such devices may also help damaged body parts to improve their function and perhaps start a person on the road to recovery. For example, some studies have found that the use of exoskeletons can aid a person with spinal cord injury or paraplegia to regain some or even all of normal body functions (see, e.g., Strickland 2016; Tamburin 2015).

Another application of robotic systems in medicine involves their use in situations in which a patient is unable to confer with a health care provider in person. For example, a person may be too elderly or physical disabled to travel to the nearest hospital to receive treatment, or an injury or other medical emergency might occur in an isolated location where high-level care is not immediately available. A number of companies have now developed electronic systems by which patient and health care giver can be connected to solve such problems. For example, a person at a nursing home might experience symptoms that suggest a serious medical problem. Using a *telemedical* system, a doctor may be able to speak directly with the patient and any medical personnel present, making a diagnosis as to the probable nature of the condition and the treatment that should be started (e.g., see "Giraff Brings People Together in the Care of Those Living at Home" 2017; "InTouch Health in Action" 2015).

Other applications of robots in medicine are still under development or in the early stages of production. An example is a device called an *origami bot* or *origami robot*, a type of miniature medical robot. This tiny device is about a centimeter in length and weighs a third of a gram; a tiny magnet is attached to its surface. When ingested, the robot unfolds and can be guided throughout the digestive system by means of external magnets. It can be used to attract and remove particles trapped in the body or to facilitate the healing of a wound. The device can also carry medications to be delivered to very specific parts of the body (Hardesty 2016).

These examples illustrate the exciting progress that has been made in the use of robotic systems in medicine, as well as even more promising advances in the future. Some concerns have been raised, however, about known or possible drawbacks from the use of some of these systems. Some of those concerns reflect the problems associated with the development and implementation of almost any new kind of technology. "Bugs" may remain in a system, resulting in less-efficient operation than will eventually be the case. Also, too few practitioners in the field may be trained in the use of medical robots to allow their widespread use. And, as with most new technologies, the costs associated with such devices can be very high, easily in the tens of thousands of dollars or more for some devices.

The bugs in a medical robotic system may have serious consequences for patients with whom they are used. A comprehensive survey of the use of the Da Vinci surgery system conducted for the period between 2000 and 2013 found a total of 10,624 adverse effects from about 1.75 million robotic surgeries in the United States. Of those adverse effects, 144 patients died, 1,391 patients were injured, and 8,061 instruments malfunctioned. The authors of the study noted that the number and type of adverse events as a result of robotic surgery had stayed relatively constant from year to year. They concluded that "adoption of advanced techniques in design and operation of robotic surgical systems and enhanced mechanisms for adverse event reporting may reduce these preventable incidents in the future" (Alenzadeh et al. 2016).

Military Applications

The use of robots for military purposes and for maintaining domestic security may be one of the oldest of all applications of robotic technology. It may well also be the most controversial of all robotic applications. Many historians point to the research of Nikola Tesla on the development of a robotic boat as the earliest example of a robot built for military purposes. Tesla had in mind military uses for his robotic boat and offered it to the United States Navy, which, however, declined the offer ("A Revolutionary Demonstration" 2017).

Research on robotic weapons stalled after Tesla's aborted efforts, becoming popular only during World War II, primarily in Germany. One of the first devices developed by German scientists was a robotic wheeled vehicle that could be directed through long cables by operators at some distance from the device. The vehicle, called The Goliath, carried about 1,000 pounds of explosives that were detonated when it approached a target building or vehicle. German researchers also were active in the development of airborne robotic devices, such as the cruise missile, V1, and ballistic missile, V2, which produced such terrible devastation in Great Britain in the last phase of the war. The United States also attempted to develop remote-controlled bombers but with much less success than for the Germans (Singer 2011).

Today, military robotics can be classified into a half dozen general categories. In many cases, devices may be available for use on the ground, in the air, and under water. The main types of military robots are discussed next. Note that many examples of each type of device are now available or in the process of development. One example only of each is given to provide a flavor of the nature of devices in each category.

- Surveillance and reconnaissance: Surveillance is the act of "keeping an eye on" some predetermined target, to see where it is located and what it is doing. Reconnaissance involves a survey of a wider area, looking for troop movements or other activities of interest to the military.
  - A recently developed example of such a device is a wireless robot whose operation is controlled via the Internet, thus providing it with an unlimited range of action. This technology represents an improvement over previous surveillance and reconnaissance vehicles that communicated with its base by means of direct electronic signals, which have a limited range of action. The vehicle has an array of sensors that allow it to detect the presence of weapons, harmful chemicals, fire, and humans. The device is self-powered, operating on solar cells that provide it with energy for extended periods of time (Kaur and Kumar 2015).

- Weapon-carrying devices: Many robots have been invented to deliver bombs and explosives to an enemy target, either under the control of a distant human or by movement determined by the robot's own computing system.

  ○ Some military robots can have more than one function. For example, the surveillance and reconnaissance device described earlier can also be equipped with weapons that allow them to fire on an enemy. Weapon-carrying robots can be either manned or unmanned and, therefore, depend on the person with whom they travel or are able to carry out their mission without the presence of a human being.

  ○ One of the most interesting of all experimental airborne robotic systems is called Perdix, a system developed by the Massachusetts Institute of Technology for the U.S. Department of Defense. The Perdix system consists of many (up to 100) small drones that can be launched from an aircraft or from ground launching stations. No single Perdix drone can accomplish its mission by itself; all Perdix drones share a common "brain" that tells them how to organize and attack a target on the ground. For a video of a Perdix "swarm" attack on a target, see "US Fighter Jets Launch Drone Swarm of Hundreds of Micro Drones: Perdix Micro-UAV Drone Swarm Test" (2017).

- Mine detection and destruction systems: According to some experts, there are about 110 million unexploded landmines in the world. On average, one person is killed every day, and about 70 more are injured by stepping on landmines. These weapons represent a threat not only to military personnel in a region of ongoing conflict but ever more so to ordinary civilians tending their fields or walking to market. For this reason, devices for detecting, marking, and disposing of landmines are among the most popular of all military robots in the world today.

- One such device is an airborne robot known as the Mine Kafon Drone. This robot operates autonomously and can carry out all three steps of a minesweeping operation. It first flies over an area where landmines are known or suspected to exist. When its sensor detects a mine, it places a marker above the mine to indicate its presence. Finally, the drone moves to a higher altitude and fires at the landmine site, causing it to explode (Antonimuthu 2016 [video]).

- Delivery system: Various types of trucks and carts have been invented to transport material and weapons to some desired location. Such devices are usually able to carry much larger cargoes than what a single person or even a group of individuals can carry.

  - Watching robotic delivery systems such as the experimental device called Big Dog is an informative introduction to the state of the art in that field. The robot is able to travel at a maximum speed of about six miles per hour, can climb up slopes of up to 35 degrees, and has a capacity of up to 70 pounds. For a demonstration of the robot in action, see About Big Dog (2017). Also see DARPA Legged Squad Support System (LS3) (2012).

- Fire-fighting devices: These robots are somewhat similar to the robots described in Chapter 1 that are able to go into environments (such as fires) that are too dangerous for humans to approach. Such devices have uses in any military situation where fires are too intense for humans to work near them. One setting in which this situation is especially difficult is on board ships at sea. In comparison with land-based fires, a ship at sea has no way of calling for additional resources to deal with an onboard fire; the ship must be ready with some type of system for dealing with the fire efficiently and quickly.

  - One such device in development is the United States Navy's Shipboard Autonomous Firefighting Robot (SAFFiR). SAFFiR consists of an android robot with sensors capable of detecting a fire and determining its most intense location. It is then capable of moving close enough to the fire to use a heavy spray of water on it. The system also has a system of "microfliers," small robots capable of moving through smoky environments and locating the source of a fire (US Navy Research 2015).

- Search and rescue robots: In any military action, some individuals become lost, are wounded, or die. Search and rescue devices are able to cover a wider area with greater precision than can humans in seeking out such individuals. These robots also have great value for nonmilitary organizations, such as police and sheriff's departments.

  ○ Some search and rescue devices have also been developed for underwater tasks. A leader in this research has been the Teledyne Marine program, which has developed a series of underwater robots known as SeaBotix robots. Models in this series can perform a variety of underwater tasks, such as surveying and studying underwater terrain; searching for objects on the ocean floor, such as airplane black boxes; delivering cargo to underwater destinations; and attaching to and examining the outer walls of sunken ships and other objects (SeaBotix Overview 2016).

The roles that robots can play in military and security actions have been applauded by many experts in those fields, along with a significant portion of the general public. But the consequences of such actions have received condemnation by others. Possibly the best single document to summarize these objections is a publication released by the Campaign to Stop Killer Robots, a program of PAX, a Netherlands-based organization dedicated to working for "a dignified, democratic and peaceful society, everywhere in the world" (Ekelhof and Struyk 2014, n.p.). Among the objections the campaign mentions in this document are the following.

- The use of robots to attack enemy forces and facilities currently depends on the use of human interventions that determine whether, when, and where such attacks occur.

That scenario in and of itself is frightening because it results in humans often at great distance from a battlefield making life and death decisions about individuals whom they will never see or with whom they will never have any personal contact. The act of killing itself is, by its very nature, an inhuman act. Therefore, depersonalizing the act with robots makes the process even more inhumane.

- The problem becomes even more serious when robots will be able, in and of themselves, to make such decisions. Lacking even the most basic moral sensibility that humans have, robots will think nothing at all of taking human life, degrading the value of life to zero among both robots and humans.

- The use of robots as killing machines also tends to remove humans' sense of responsibility for the loss of life. As humans remain at distance from the actually killing process, or when they are removed completely from the process, the act of killing is likely to be viewed less and less as a moral issue and more of a tactical decision for ways of ending a conflict.

- The precision of robotic weapons is also a matter of concern. At the present stage of development, it is difficult or impossible for most such weapons to distinguish between legitimate targets, such as a military command headquarters, and a civilian facility, such as a hospital, or between enemy combatants and civilians. As a consequence, the amount of collateral damage, harm to individuals and facilities not involved in a conflict, is likely to be high with robotic weapons. The precision with which these devices operate is likely to improve, but it seems unlikely that they will ever achieve the level of distinction between "friend" and "foe" that humans have, thus violating one of the basic principles of international law ("Rule 1. The Principle of Distinction between Civilians and Combatants" 2017).

- A related objection is based on the principle of *proportionality* in armed conflict. According to that principle, the damage caused on a civilian population should not be greater than the benefits of the military action. Opponents of robotic warfare worry that fully automated weapons (FAWs) would be inclined to go "all out" in an attack, obliterating both enemy forces and the civilian population ("Making the Case: The Dangers of Killer Robots and the Need for a Preemptive Ban" 2016).

- Critics of robotic warfare also point to the problem of *accountability* in such actions. Whatever the results of an FAW attack—success in destroying the enemy, collateral damage among civilians, or any other result—the question remains as to who is responsible for those results. If a children's hospital is destroyed in a robotic attack, who is to blame for that error: the robot itself (whatever it means to assign "responsibility" to the robot), the military commander, the manufacturer of the robot, the governmental agency that authorized the use of robotic warfare in this instance? The concern is that, as in so many other cases, if accountability cannot be assigned, it may end up that no one is responsible for the action (Guiora 2017).

- Proliferation of robotic weapons is yet another concern for opponents of the technology. Military historians point out that new and more effective weapons technologies do not remain the sole possession of the one country where they were developed. Over time, other countries learn about, research, and eventually develop the technology, generally in newer and more powerful forms. The development of nuclear weapons after World War II is an example of such a process. The future of robotic warfare would appear to hold the potential for a similar process, with each country trying to develop new and better robots for the battlefield, resulting in an escalation of military power that could have devastating worldwide consequences. Proliferation to nonmilitary entities, such as national, regional, and local law enforcement agencies, is a comparable concern (see, e.g., Horowitz and Fuhrmann 2017).

For each of the arguments made earlier, there is a contrary argument in support of the expanded use of FAW systems. This debate is likely to continue into the near-future, at least, and probably much longer (Etzioni and Etzioni 2017).

Space Research

When was the first space robot launched? That might seem to be a simple question, but there is no simple answer. Some authorities cite a suborbital flight by the Soviet Union in 1951 that carried two dogs into space for this honor (Siddiqi 2003, 96). Others point to the launch of the world's first artificial satellite, Sputnik I, on October 4, 1957, again by the Soviet Union. Many other robotic "firsts" can be listed, such as the first U.S. satellite launched into space, the first device to travel around and/or to the surface of the moon, the first space probe to visit another planet, or the first rover to travel to the moon or another planet ("What Is Robotics?" 2009). In any case, robots—whether defined as "machines that can be used to do jobs," "reprogrammable, multifunctional manipulator[s] designed to move material, parts, tools, or specialized devices through variable programmed motions for the performance of a variety of tasks," or "one-armed, blind idiot[s] with limited memory and which cannot speak, see, or hear" ("Robotics" 2017, 3)—have been around for at least 60 years. They include thousands of specific autonomous or semiautonomous devices for sending information about the solar system to the Earth and carrying out other useful functions. (Probably the most complete list of all space flights can be found at "Today in Space History" 2017.)

For many people, the term *space robot* may actually bring to mind a *space android*, a robot that looks and acts very much like humans. Science fiction writers have sometimes described the activities of such machines on Earth, on another planet, or in a free-flying spacecraft. By this definition, the first operable "space robot" in the world was released by NASA in 2010. The robot, known as Robonaut 2, was sent to the International

Space Station, where it performs some of the tasks otherwise carried out by human astronauts at the station ("Robonaut 2— NASA's Humanoid Robot" 2013).

As a practical matter, the vast majority of space robots in use today or under development are devices that do not look anything like humans but that are still capable of carrying out many essential functions in space exploration. The Mars rovers discussed in Chapter 1 are examples of such devices. The question to be answered in the future of space exploration is whether space robots can replace humans in *all* aspects of space travel, or whether there are some missions that can be performed only by humans. Debates over the future of the exploration of Mars are perhaps the most immediate of those discussions.

The case for robots, rather than humans, as space explorers is based on a number of arguments. Probably the most common and most important of these arguments is that developing robotic technology for the exploration of the solar system is far less expensive than finding a way to send humans to another planet or solar system body. NASA's most recent estimates put the cost of developing a human-to-Mars program to be between $80 billion and $100 billion, with a development period of about 20 years. By comparison, the most recent robotic program, Orion, is estimated to cost a total of less than $4 billion and is expected to launch in 2019 (Kaufman 2014).

A large part of that cost, and a second reason for humans *not* to undertake such space travel, is the unfriendly environment through which human astronauts would have to travel and under which they would have to survive once on another solar body. The time it would take just to get to another body beyond the moon might also prove to be prohibitive (Gonzalez 2016; for an excellent and more detailed discussion of this point of view, see Dvorsky 2014).

Those who advocate for human space travel point out that, as smart as existing robots may be, they can never achieve the skills associated with human intelligence: being able to respond to unexpected events, analyzing data to understand events, and deciding on new pathways in a research project. Perhaps an even more cogent argument is that the future of humans is in space. The species cannot survive on this planet forever in its present form. Humans will eventually have to explore other planets and solar bodies to find places to which human civilization can expand. In that context, space exploration by humans is not really a choice; it is a necessity. One of the best-known spokespersons for this position is the eminent theoretical physicist Stephen Hawking. In 2017, Hawking made a documentary film, "Expedition New Earth," in which he argues that humans will have to find another planet to colonize within 100 years, or the species will become extinct on this planet. He points to the threats facing the human species, such as climate change, possible asteroid collisions, overpopulation, and epidemics to justify his warning ("Stephen Hawking Warns We Have 100 Years to Leave Earth" 2017 [video]).

Learning about Robots

Robots have become an integral part of human societies. In recognizing that fact, many educational institutions are now offering formal courses and programs in robotics. Some of the most popular of these courses have been the *robot camps* operated by schools, colleges, universities, and educational corporations. In 2017, more than 100 robot camps were offered in the United States for children and young adults aged 5–19 ("Summer Camps" 2017). Instruction in most of these camps focuses on learning how to build and program a robot. Probably the most common tool in these programs are Lego blocks, invented by Danish carpenter Ole Kirk Christiansen in 1932. Originally designed as a wooden toy with which children could build a variety of structures, today they are made of plastics and have instructional value in addition to still being playthings for children and young adults. (The Lego Web site is https://www .lego.com/en-us/.)

Another mechanism for teaching about robots has been competitions among teams of children and young adults. These competitions are designed to encourage research by participants to make the most efficient robot for the completion of some task or set of tasks. Probably the largest such competition in the world today is the FIRST (For Inspiration and Recognition of Science and Technology) founded by inventor Dean Kamen and physicist Woodie Flowers in 1989. The goal of the program has long been to encourage young people to become more interested in and informed about robotic technology. In addition to national and worldwide contests, FIRST sponsors robot leagues for children in grades K-4 and 4–8, as well as the FIRST Technical Challenge for middle and high school students and the international FIRST Robotics Competition for students in grades 9–12. The first FIRST program was a contest among 28 high school teams held in a New Hampshire high school gymnasium in 1992. The most recent FIRST Robotics Competition was held in Washington, D.C., on July 16–18, 2017, at which 163 nations and regions were represented, ranging from Afghanistan and Sierra Leone to Cambodia and Vanuatu ("Team Alphabot Makes Pakistan Proud @ the FIRST Global Challenge" 2017; also see Chapter 3 for an essay from the last of these teams).

Conclusion

From toy, religious symbol, and source of entertainment, robots and automata have evolved to become the workhorses of modern society on which humans depend for a variety of tasks. This change has inspired excitement among many individuals because of the further benefits robots can bring to human civilization. But it has also raised some profound questions about the proper role of robots in society, as well as even more basic questions about the nature of the human species and its future in a world filled with robotic devices.

References

"About." 2017. Campaign against Sex Robots. https://campaign againstsexrobots.org/about/. Accessed on September 8, 2017.

"About Big Dog." 2017. Boston Dynamics. https://www .bostondynamics.com/bigdog. Accessed on September 12, 2017.

"About FHI." 2017. The Future of Humanity Institute. https://www.fhi.ox.ac.uk/about/about-fhi/. Accessed on August 26, 2017.

"Aethon TUG Robot." 2015. YouTube. https://www.youtube .com/watch?v=kCDJObCNufg. Accessed on September 8, 2017.

"AI Robot Sophia and Her Plans to 'Dominate the Human Race.'" 2017. YouTube. https://www.youtube.com/ watch?v=-bXUyTiMHB4. Accessed on March 16, 2018.

Alenzadeh, Homa, et al. 2016. "Adverse Events in Robotic Surgery: A Retrospective Study of 14 Years of FDA Data." *PLoS One* 11(4): e0151470. PMCID: PMC4838256. https://www.ncbi.nlm.nih.gov/pmc/articles/PMC483 8256/. Accessed on September 11, 2017.

"Aliens, Very Strange Universes and Brexit—Martin Rees Q&A." 2017. The Conversation. https://theconversation .com/aliens-very-strange-universes-and-brexit-martin-rees-qanda-75277. Accessed on August 25, 2017.

"All about Robotic Surgery." 2018. AVRA Medical Robots, Inc. http://allaboutroboticsurgery.com/booksonmedical robotics.html. Accessed on March 15, 2018.

Allen, Ben. 2015. "How Robots Are Transforming Healthcare." ZME Science. http://www.zmescience.com/ medicine/robots-and-healthcare-0423/. Accessed on September 8, 2017.

Allen, Colin. 2011. "The Future of Moral Machines." *New York Times*. https://opinionator.blogs.nytimes.com/2011/

12/25/the-future-of-moral-machines/?mcubz=3&_r=0.
Accessed on September 3, 2017.

Allen, Colin, Gary Varner, and Jason Zinser. 2000.
"Prolegomena to Any Future Artificial Moral Agent."
Journal of Experimental & Theoretical Artificial Intelligence
12(3): 251–261. https://www.researchgate.net/
publication/220080115_Prolegomena_to_any_future_
artificial_moral_agent. Accessed on August 31, 2017.

Allen, Paul G. 2011. "Paul Allen: The Singularity Isn't Near."
MIT Technology Review. https://www.technologyreview
.com/s/425733/paul-allen-the-singularity-isnt-near/.
Accessed on August 27, 2017.

Amodei, Dario, et al. 2016. "Concrete Problems in AI Safety."
arXiv:1606.06565v2. https://arxiv.org/pdf/1606.06565v2
.pdf. Accessed on September 4, 2017.

Anderson, Michael, Susan Leigh Anderson, and Chris
Armen. 2006. "An Approach to Computing Ethics." *IEEE
Intelligent Systems* 21(4): 56–63. https://ieet.org/archive/
IEEE-Anderson.pdf. Accessed on September 2, 2017.

Antonimuthu, Rajamanickam. 2016. "Mine Kafon Drone—
An Airborne Demining System to Clear Land Mines
around the World in 10 Years." YouTube. https://www
.youtube.com/watch?v=uCkkuBAx4c4. Accessed on
September 12, 2017.

Arnold, Thomas, and Matthias Scheutz. 2016. "Against the
Moral Turing Test: Accountable Design and the Moral
Reasoning of Autonomous Systems." *Ethics and Information
Technology* 18(2): 103–115.

"Artificial Intelligence and Life in 2030." 2016. https://ai100
.stanford.edu/sites/default/files/ai100report10032016fnl_
singles.pdf. Accessed on August 27, 2016.

"Asilomar Study on Long-Term AI Futures." 2009. American
Association for the Advancement of Science. http://

erichorvitz.com/AAAI_Asilomar_Study.pdf. Accessed on March 15, 2018.

Asimov, Isaac. 1942. "Runaround." http://web.williams.edu/ Mathematics/sjmiller/public_html/105Sp10/handouts/ Runaround.html. Accessed on August 24, 2017.

Barden, Jessica. 2017. "The Pros and Cons of Having Robots in the Workplace." RecruitLoop. http://recruitloop.com/ blog/the-pros-and-cons-of-having-robots-in-the-work place/. Accessed on September 5, 2017.

Berg, Andrew, Edward F. Buffie, and Luis-Felipe Zanna. 2016. "Robots, Growth, and Inequality." *Finance & Development* 53(3) 10–13. http://www.imf.org/external/pubs/ft/ fandd/2016/09/berg.htm. Accessed on August 29, 2017.

Bers, Marina U., et al. 2017. DevTech Research Group. Tufts University. http://ase.tufts.edu/devtech/tangiblek/ Classroom%20Curriculum%20Version%201.02%20 Nov%208%202010.pdf. Accessed on September 7, 2017.

Brownlee, John. 2016. "Google Created Its Own Laws of Robotics." CO.Design. https://www.fastcodesign.com/ 3061230/google-created-its-own-laws-of-robotics. Accessed on September 4, 2017.

Butler, Jason. 2016. "Robots Can't Solve Real-World Financial Problems." *Financial Times*. https://www.ft.com/content/ 9976553a-b586-11e6-961e-a1acd97f622d. Accessed on September 6, 2017.

Carlopio, Jim. 1988. "A History of Social Psychological Reactions to New Technology." *Journal of Occupational and Organizational Psychology* 61(1): 67–77.

Cellan-Jones, Rory. 2014. "Stephen Hawking Warns Artificial Intelligence Could End Mankind." BBC News. http:// www.bbc.com/news/technology-30290540. Accessed on August 25, 2017.

Christian, Brian. 2011. "Mind vs. Machine." *Atlantic.* https://www.theatlantic.com/magazine/archive/2011/03/mind-vs-machine/308386/. Accessed on August 28, 2017.

"Collision and Injury Criteria When Working with Collaborative Robots." 2012. Health and Safety Group. http://www.hse.gov.uk/research/rrpdf/rr906.pdf. Accessed on September 5, 2017.

Crews, John. 2016. *Robonomics: Prepare Today for the Jobless Economy of Tomorrow.* North Charlestown, SC: CreateSpace.

"DARPA Legged Squad Support System (LS3)." 2012. DARPAtv. YouTube. https://www.youtube.com/watch?v=xY42w1w0TWk. Accessed on September 12, 2017.

Dashevsky, Evan. 2014. "20 Jobs Likely to Be Replaced by Robots (and 20 That Are Safe)." *PC Magazine.* https://www.pcmag.com/article2/0,2817,2459986,00.asp. Accessed on August 28, 2017.

Delaney, Kevin J. 2017. "Why Bill Gates Would Tax Robots." Quartz. https://qz.com/911968/bill-gates-the-robot-that-takes-your-job-should-pay-taxes/. Accessed on August 29, 2017.

Delsen, Lei, and Jeroen Smits. 2014. "The Rise and Fall of the Dutch Savings Schemes." Network for Studies on Pensions, Aging and Retirement. http://arno.uvt.nl/show.cgi?fid=133632. Accessed on August 30, 2017.

"Do We Need Asimov's Laws?" 2014. *Technology Review.* https://www.technologyreview.com/s/527336/do-we-need-asimovs-laws/#comments. Accessed on September 4, 2017.

Dowd, Maureen. 2017. "Elon Musk's Billion-dollar Crusade to Stop the A.I. Apocalypse." *Vanity Fair.* https://www.vanityfair.com/news/2017/03/elon-musk-billion-dollar-crusade-to-stop-ai-space-x. Accessed on August 25, 2017.

Dvorsky, George. 2014. "12 Reasons Robots Will Always Have an Advantage over Humans." Gizmodo. http://io9

.gizmodo.com/12-reasons-robots-will-always-have-an-advantage-over-hu-1671721194. Accessed on September 5, 2017.

Ekelhof, Merel, and Miriam Struyk. 2014. "Deadly Decisions: 8 Objections to Killer Robots." https://www.paxforpeace.nl/publications/all-publications/deadly-decisions. Accessed on September 13, 2017.

Etzioni, Amitai, and Oren Etzioni. 2017. "Pros and Cons of Autonomous Weapons Systems." *Military Review.* http://www.armyupress.army.mil/Portals/7/military-review/Archives/English/pros-and-cons-of-autonomous-weapons-systems.pdf. Accessed on September 13, 2017.

Etzioni, Oren. 2017. "How to Regulate Artificial Intelligence." *New York Times.* https://www.nytimes.com/2017/09/01/opinion/artificial-intelligence-regulations-rules.html?mcubz=3&_r=0. Accessed on September 4, 2017.

"European Parliament Calls for Robot Law, Rejects Robot Tax." 2017. Reuters. http://www.reuters.com/article/us-europe-robots-lawmaking-idUSKBN15V2KM. Accessed on August 29, 2017.

"FAQ (Frequently Asked Questions)." 2017. TrueCompanion.com. http://www.truecompanion.com/shop/faq. Accessed on September 7, 2017.

Floyd, Patricia A., and Beverly Joyce Allen. 2009. *Introduction to Careers in Health, Physical Education, and Sport.* Belmont, CA: Wadsworth Cengage Learning.

Ford, Martin. 2015. *Rise of the Robots: Technology and the Threat of a Jobless Future.* New York: Basic Books.

"Giraff Brings People Together in the Care of Those Living at Home." 2017. Giraff. http://www.giraff.org/?lang=en. Accessed on September 10, 2017.

Gonzalez, Rowan. 2016. "Why Robots Are Better Space Explorers." Computer Stories. https://computerstories.net/why-robots-are-better-space-explorers-17985. Accessed on September 14, 2017.

Gregory, Caroline. 2014. "A Field Guide to Anti-Technology Movements, Past and Present." *HuffPost*. http://www .huffingtonpost.com/2014/01/17/life-without-technology-t_n_4561571.html. Accessed on August 28, 2017.

Groopman, Jerome. 2009. "Robots That Care." *New Yorker*. http://www.newyorker.com/magazine/2009/11/02/robots-that-care. Accessed on September 7, 2017.

Guiora, Amos N. 2017. "Accountability and Decision Making in Autonomous Warfare: Who Is Responsible?" *Utah Law Review* 2(4): 393–422. http://dc.law.utah.edu/ cgi/viewcontent.cgi?article=1047&context=ulr. Accessed on September 13, 2017.

Hardesty, Larry. 2016. "Ingestible Origami Robot." MIT News. http://news.mit.edu/2016/ingestible-origami-robot-0512. Accessed on September 9, 2017.

Hawkins, Jeff. 2015. "The Terminator Is Not Coming. The Future Will Thank Us." Recode. https://www.recode.net/ 2015/3/2/11559576/the-terminator-is-not-coming-the-future-will-thank-us. Accessed on August 27, 2017.

Horowitz, Michael C., and Matthew Fuhrmann, 2017. "Droning On: Explaining the Proliferation of Unmanned Aerial Vehicles." *International Organization* 71(2): 397–418. https://poseidon01.ssrn.com/delivery.php?ID=22711 11110640750020041120970650740860970340610550016 02002710209509912000609401006909603000002010030 07009008123077022088070020091051007069048051082 07211411509408910210506909208503008106500300906 50800680011140950650010941080200690080160050 75096005123116098&EXT=pdf. Accessed on September 13, 2017.

"Individual Activity Accounts Germany." 2017. http://www .oecd.org/els/soc/20_Blancke_Individual_Accounts.pdf. Accessed on August 30, 2017.

"Innovations in Finance and Risk—Robotic Process Automation." 2016. Accenture. https://www.slideshare.net/

accenture/applying-robotic-process-automation-in-banking-innovations-in-finance-and-risk. Accessed on September 6, 2017 (slide show).

"InTouch Health in Action." 2015. YouTube. https://www.youtube.com/watch?v=gUfxgwJfq3M. Accessed on September 10, 2017.

Kaufman, Marc. 2014. "A Mars Mission for Budget Travelers." National Geographic. http://news.national geographic.com/news/2014/04/140422-mars-mission-manned-cost-science-space/. Accessed on September 14, 2017.

Kaur, Tarunpreet, and Dilip Kumar. 2015. "Wireless Multifunctional Robot for Military Applications." Second International Conference on Recent Advances in Engineering & Computational Sciences. December 21–22, 2015, Chandigarh, India. https://www.researchgate.net/publication/304239400_Wireless_multifunctional_robot_for_military_applications. Accessed on September 12, 2017.

Kingma, Luke. 2017. "Universal Basic Income: The Answer to Automation?" Futurism. https://futurism.com/images/universal-basic-income-answer-automation/. Accessed on September 1, 2017.

Kleeman, Jenny. 2017. "The Race to Build the World's First Sex Robot." *Guardian.* https://www.theguardian.com/technology/2017/apr/27/race-to-build-world-first-sex-robot. Accessed on September 7, 2017.

Knapton, Sarah. 2016. "Robots Will Take Over Most Jobs within 30 Years, Experts Warn." *Telegraph.* http://www.telegraph.co.uk/news/science/science-news/12155808/Robots-will-take-over-most-jobs-within-30-years-experts-warn.html. Accessed on August 28, 2017.

"Kohlberg's Stages of Moral Development." 2017. Psychology Charts. http://www.psychologycharts.com/kohlberg-stages-of-moral-development.html. Accessed on September 3, 2017.

Kosoff, Maya. 2017. " 'We Do Not Have Long': Elon Musk Calls for a Global Ban on Killer Robots." *Vanity Fair.* https://www.vanityfair.com/news/2017/08/elon-musk-calls-for-a-global-ban-on-killer-robots. Accessed on August 25, 2017.

Koteswara Karthik, P., and P. Ravi Chandra. 2014. "An Overview of Agricultural Robots." *Yuva Engineers.* http://www.yuvaengineers.com/an-overview-of-agricultural-robots-p-koteswara-karthik-p-ravi-chandra. Accessed on September 5, 2017.

Krebs, H. I., and B. T. Volpe. 2013. "Rehabilitation Robotics." *Handbook of Clinical Neurology* 110: 283–294. https://www.ncbi.nlm.nih.gov/pmc/articles/PMC4688009/. Accessed on September 10, 2017.

Kurzweil, Ray. 2005. *The Singularity Is Near: When Humans Transcend Biology.* New York: Penguin Books.

Lam, Tin Lun, and Yangsheng Xu. 2011. "A Flexible Tree Climbing Robot: Treebot—Design and Implementation." 2011 IEEE International Conference on Robotics. Abstract and images at https://www.semanticscholar.org/paper/A-flexible-tree-climbing-robot-Treebot-design-and-Lam-Xu/638329568c1a305e52114591276a80face9309df. Accessed on September 2017.

Littman, Michael. 2015. " 'Rise of the Machines' Is Not a Likely Future." Live Science. https://www.livescience.com/49625-robots-will-not-conquer-humanity.html. Accessed on August 26, 2017.

Lodhi, Hamayal Wajid, Aleena Ahmed Khan, and Maria Aziz. 2013. "Agricultural Robots." In Slide Share. https://www.slideshare.net/linokhan/agricultural-robot. Accessed on September 5, 2017.

López, Leticia Lafuente. 2016. "Educational Robots at Global Robot Expo." eLearning Industry. https://elearningindustry.com/educational-robots-global-robot-expo. Accessed on September 7, 2017.

Lucaciu, Irina. 2013. "Robots: The Answer for Treating Children with Autism Spectrum Disorder?" The Neuroethics Blog. http://www.theneuroethicsblog.com/ 2013/07/robots-answer-for-treating-children.html? utm_campaign=elearningindustry.com&utm_source= %2Feducational-robots-global-robot-expo&utm_ medium=link. Accessed on September 7, 2017.

"Making the Case: The Dangers of Killer Robots and the Need for a Preemptive Ban." 2016. Human Rights Watch. https://www.hrw.org/report/2016/12/09/making-case/ dangers-killer-robots-and-need-preemptive-ban. Accessed on September 13, 2017.

Manyika, James, et al. 2017. "A Future That Works: Automation, Employment, and Productivity." http:// www.mckinsey.com/global-themes/digital-disruption/ harnessing-automation-for-a-future-that-works. Accessed on August 29, 2017.

Marinov, Bobby. 2016. "Medical Exoskeletons into 6 Categories." Exoskeleton Report. http://exoskeletonreport .com/2016/06/medical-exoskeletons/. Accessed on September 10, 2017.

McDonald, Coby. 2015. "The Good, the Bad and the Robot: Experts Are Trying to Make Machines Be 'Moral.' " 2015. Cal Alumni Association. https://alumni.berkeley.edu/ california-magazine/just-in/2015-06-08/good-bad-and- robot-experts-are-trying-make-machines-be-moral. Accessed on September 3, 2017.

McNeal, Marguerite. 2017. "Rise of the Machines: The Future Has Lots of Robots, Few Jobs for Humans." *Wired*. https://www.wired.com/brandlab/2015/04/rise-machines- future-lots-robots-jobs-humans/. Accessed on August 28, 2017.

"Moral Computations." 2017. Human Robot Interaction Laboratory. Tufts University. https://hrilab.tufts.edu/publi

cations/?researchFilter=moral&authorFilter=&yearFilter=. Accessed on August 31, 2017.

Morgan, Jacob. 2016. "What Is the Fourth Industrial Revolution?" *Forbes*. https://www.forbes.com/sites/jacobmorgan/2016/02/19/what-is-the-4th-industrial-revolution/#b2795d7f392a. Accessed on August 29, 2017.

Murphy, Robin R., and David D. Woods. 2009. "Beyond Asimov: The Three Laws of Responsible Robotics." *IEEE Intelligent Systems* 24(4): 14–20. http://www.inf.ufrgs.br/~prestes/Courses/Robotics/beyond%20asimov.pdf. Accessed on August 24, 2017.

Nisen, Max. 2013. "Robot Economy Could Cause up to 75 Percent Unemployment." *Business Insider*. http://www.businessinsider.com/50-percent-unemployment-robot-economy-2013-1. Accessed on August 28, 2017.

Obias, Rudy. 2015. "10 Therapy Robots Designed to Help Humans." Mental Floss. http://mentalfloss.com/article/71987/10-therapy-robots-designed-help-humans. Accessed on September 7, 2017.

"An Open Letter: Research Priorities for Robust and Beneficial Artificial Intelligence." 2015. Future of Life Institute. https://futureoflife.org/ai-open-letter. Accessed on August 26, 2017.

OpenAI. 2017. https://openai.com/. Accessed on August 25, 2017.

Palocsik, Jenni. 2016. "25 Examples of Processes for Robotic Process Automation." Verint. http://blog.verint.com/customer-engagement/25-examples-of-processes-for-robotic-process-automation. Accessed on September 6, 2017.

Pepe, Courtney. 2016. "How to Use Robots in the Elementary Classroom." *Daily Genius*. http://dailygenius.com/robots-classroom/. Accessed on September 7, 2017.

Percy, Nathan. 2017. "Santa Margarita High Teacher Uses Robot to Teach Class from China during Teen Tech Week." *Orange County Register*. http://www.ocregister.com/2017/ 03/09/santa-margarita-high-teacher-uses-robot-to-teach-class-from-china-during-teen-tech-week/. Accessed on September 7, 2017.

Perry, Tekla S. 2013. "Profile: Veebot. Making a Robot That Can Draw Blood Faster and More Safely Than a Human Can." *IEEE Spectrum*. https://spectrum.ieee.org/robotics/medical-robots/profile-veebot. Accessed on September 9, 2017.

"The Personal Activity Account." 2017. Gouvernement.fr. http://www.gouvernement.fr/compte-personnel-activite-cpa (translatable into English). Accessed on August 30, 2017.

Pilarski, Thomas, et al. 2002. "The Demeter System for Automated Harvesting." *Autonomous Robots* 13(1): 9–20. http://www.frc.ri.cmu.edu/~axs/doc/ans99.pdf. Accessed on September 5, 2017.

Rainie, Lee, and Janna Anderson. 2017. "The Future of Jobs and Jobs Training." Pew Research Center. http://www.pewinternet.org/2017/05/03/the-future-of-jobs-and-jobs-training/. Accessed on August 30, 2017.

"Remote Student." 2013. VGo Communications. http:// www.vgocom.com/remote-student. Accessed on September 7, 2017.

"A Revolutionary Demonstration." 2017. PBS. https:// www.pbs.org/tesla/ins/lab_remotec.html. Accessed on September 11, 2017.

Richardson, Kathleen. 2015. "The Asymmetrical 'Relationship': Parallels between Prostitution and the Development of Sex Robots." Campaign against Sex Robots. *SIGCAS Computers & Society* 45(3): 290–293. https://campaignagainstsexrobots.org/the-asymmetrical-relationship-parallels-between-prostitution-and-the-development-of-sex-robots/. Accessed on September 8, 2017.

"Robonaut 2—NASA's Humanoid Robot." 2013. YouTube. https://www.youtube.com/watch?annotation_id=annota tion_1062681065&feature=iv&src_vid=g3u48T4Vx7k& v=ePWjFlSdB4U. Accessed on September 14, 2017.

"Robot Ethics: Morals and the Machine." 2012. *Economist*. 8787: 13. http://www.economist.com/node/21556234. Accessed on September 4, 2017.

"Robot Pets Offer Real Comfort." 2016. CNN. http://www .cnn.com/2016/10/03/health/robot-pets-loneliness/index .html. Accessed on September 7, 2017.

"Robotic Surgery Demonstration Using Da Vinci Surgical System." 2012. Future Trends. YouTube. https://www.you tube.com/watch?v=VJ_3GJNz4fg. Accessed on September 9, 2017.

"Robotic Syringes." 2017. Robots and Androids. http://www .robots-and-androids.com/robotic-syringes.html. Accessed on September 9, 2017.

"Robotics." 2017. Engineering Innovation. Johns Hopkins Whiting School of Engineering. https://engineering .jhu.edu/ei/wp-content/uploads/sites/29/2014/01/ Robotics-PowerPoint1.pdf. Accessed on September 14, 2017.

"Robots: What Are the Risks?" 2016. *Zurich Insider*. https:// insider.zurich.co.uk/industry-spotlight/robots-what-are- the-risks/. Accessed on September 5, 2017.

Rosenbaum, Eric. 2017. "IBM's Watson Supercomputer Is Getting into Wall Street Stock-Picking." CNBC. https:// www.cnbc.com/2017/06/16/ai-assault-on-stock-market- ibms-watson-is-getting-into-etf-business.html. Accessed on September 6, 2017.

"Rule 1. The Principle of Distinction between Civilians and Combatants." 2017. Customary IHL. https://ihl-databases .icrc.org/customary-ihl/eng/docs/v1_rul_rule1. Accessed on September 13, 2017.

Sale, Kirkpatrick. 1996. *Rebels against the Future: The Luddites and Their War on the Industrial Revolution: Lessons for the Computer Age.* London: Quartet Books.

Schwab, Klaus. 2016. "The Fourth Industrial Revolution." 2016. World Economic Forum. https://www.weforum.org/agenda/2016/01/the-fourth-industrial-revolution-what-it-means-and-how-to-respond/. Accessed on March 16, 2018.

"SeaBotix Overview." 2016. Teledyne Marine Vehicles. YouTube. https://www.youtube.com/watch?v=Pt45ofDjXMk. Accessed on September 12, 2017.

"17 Definitions of the Technological Singularity." 2012. Singularity. https://www.singularityweblog.com/17-definitions-of-the-technological-singularity/. Accessed on August 26, 2017.

Sharkey, Noel, et al. 2017. "Our Sexual Future with Robots." Foundation for Responsible Robotics. https://robot republic.org/sex-with-robots/. Accessed on March 16, 2018.

Siddiqi, Asif. 2003. *Sputnik and the Soviet Space Challenge.* Gainesville: University Press of Florida.

Singer, P. W. 2011. "Drones Don't Die—A History of Military Robots." HistoryNet. http://www.historynet.com/drones-dont-die-a-history-of-military-robotics.htm. Accessed on March 16, 2018.

"Social Benefits." 2001. Glossary of Statistical Terms. Organisation for Economic Co-Operation and Development. https://stats.oecd.org/glossary/detail.asp?ID=2480. Accessed on August 30, 2017.

Steinberg, Alexander. 2016. "Over 20 Benefits of Robotic Process Automation (RPA)." LinkedIn. https://www.linkedin.com/pulse/over-20-benefits-robotic-process-auto mation-rpa-alex-steinberg. Accessed on September 6, 2017.

"Stephen Hawking Warns We Have 100 Years to Leave Earth." 2017. The Last News. https://www.youtube.com/watch?v=2iJD5LmjJo4. Accessed on September 14, 2017.

Strickland, Eliza. 2016. "Demo: The Ekso GT Robotic Exoskeleton for Paraplegics and Stroke Patients." *IEEE Spectrum*. https://spectrum.ieee.org/the-human-os/ biomedical/bionics/paraplegic-man-walks-in-ekso-robotic-exoskeleton-to-demo-its-killer-app. Accessed on September 10, 2017.

Strother, Jason. 2011. "South Korean Students Learn English from Robot Teacher." VOA. https://www.voanews.com/a/ south-korean-students-learn-english-from-robot-teacher-117640783/167151.html. Accessed on September 7, 2017.

Sullins, John P. 2006. "When Is a Robot a Moral Agent." *International Review of Information Ethics* 6(12): 23–30. http://www.realtechsupport.org/UB/WBR/texts/Sullins_ RobotMoralAgent_2006.pdf. Accessed on September 3, 2017.

"Summer Camps." 2017. NASA. https://robotics.nasa.gov/ students/summer_camps.php. Accessed on September 15, 2017.

Tamburin, Adam. 2015. "Vanderbilt Exoskeleton Helps Paralyzed Student Walk." *Tennessean*. http://www.tennes sean.com/story/news/education/2015/05/12/vanderbilt-exoskeleton-helps-paralyzed-student-walk/27206553/. Accessed on September 10, 2017.

"Team Alphabot Makes Pakistan Proud @ the First GLOBAL Challenge." 2017 FIRST Global. http://first.global/in-the-news/. Accessed on September 15, 2017.

Thompson, Derek. 2015. "A World without Work." *Atlantic*. https://www.theatlantic.com/magazine/archive/2015/07/ world-without-work/395294/. Accessed on August 31, 2017.

Thornton, R. 1847. "The Age of Machinery." *Primitive Expounder*. August 12, 1847. https://books.google.com/ books?id=ZM_hAAAAMAAJ&pg=PA281&lpg=PA281 &dq=%22such+machines,+by+which+the+scholar+may,

+by+turning+a+crank,+grind+out+the+solution+of+a+ problem%22+%22common+school+advocate%22&sou rce=bl&ots=4dV3qXUUcW&sig=aUFg_CI7R7zQSceu O8hHJ6CpRCM&hl=en&sa=X&ved=0ahUKEwi3zLX BqPXVAhVBQSYKHZIbCocQ6AEIKDAA#v=onepage &q=%22such%20machines%2C%20by%20which%20 the%20scholar%20may%2C%20by%20turning%20a%20 crank%2C%20grind%20out%20the%20solution%20of %20a%20problem%22%20%22common%20school%20 advocate%22&f=false. Accessed on August 26, 2017.

Timmins, Beth. 2017. "New Sex Robots with 'Frigid' Setting to Allow Men to Simulate Rape." *Independent.* http://www .independent.co.uk/life-style/sex-robots-frigid-settings- rape-simulation-men-sexual-assault-a7847296.html. Accessed on September 8, 2017.

"Today in Space History." 2017. http://www.astronautix.com/. Accessed on September 14, 2017.

"21 Jobs Where Robots Are Already Replacing Humans." 2016. MSN Money. http://www.msn.com/en-us/money/ careersandeducation/21-jobs-where-robots-are-already- replacing-humans/ss-BBv6yiU. Accessed on August 28, 2017.

"US Fighter Jets Launch Drone Swarm of Hundreds of Micro Drones: Perdix Micro-UAV Drone Swarm Test." 2017. WarLeaks—Daily Military Defense Videos & Combat Footage. YouTube. https://www.youtube.com/ watch?v=5NGgHyfPGU0. Accessed on September 12, 2017.

US Navy Research. 2015. "Shipboard Autonomous Firefighting Robot—SAFFiR." YouTube. https://www .youtube.com/watch?v=K4OtS534oYU. Accessed on September 12, 2017.

Vukobratovic, Miomir K. 2007. "When Were Active Exoskeletons Actually Born?" *International Journal of*

Humanoid Robotics 4(3): 459–486. http://www.pupin.rs/
RnDProfile/pdf/exoskeletons.pdf. Accessed on September
10, 2017.

Wallach, Wendell, and Colin Allen. 2010. *Moral Machines:
Teaching Robots Right from Wrong.* New York: Oxford
University Press.

Wasic, John. 2016. "How the GOP Social Security/Medicare
Doomsday Machine Works." *Forbes.* https://www.forbes
.com/sites/johnwasik/2016/12/09/how-the-gop-social-
securitymedicare-doomsday-machine-works/#5b9a37ae
2965. Accessed on August 30, 2017.

Watters, Audrey. 2015. "The First Teaching Machines." Hack
Education. http://hackeducation.com/2015/02/03/the-
first-teaching-machines. Accessed on September 6, 2017.

Weir, Kirsten. 2015. "Robo Therapy." *Monitor on Psychology.*
http://www.apa.org/monitor/2015/06/index.aspx. Accessed
on September 7, 2017.

Weiss, Suzannah. 2017. "Why Are Sex Robots All Female?
Because the 'Ideal Woman' Is a Robot." *Glamour.*
https://www.glamour.com/story/sex-robots. Accessed on
September 7, 2017.

West, Darrell M. 2015. "What Happens If Robots Take
the Jobs? The Impact of Emerging Technologies on
Employment and Public Policy." Center for Technology
Innovation at Brookings. https://www.brookings.edu/
wp-content/uploads/2016/06/robotwork.pdf. Accessed
on August 29, 2017.

"What Did It Look Like?" 2013. Slide Share. https://www
.slideshare.net/janetpareja/industrial-revolution-pp-2013-
47115535. Accessed on August 28, 2017.

"What Is Robotics?" 2009. NASA. https://www.nasa.gov/
audience/forstudents/k-4/stories/nasa-knows/what_is_
robotics_k4.html. Accessed on September 14, 2017.

Wright, Paul J., and Robert S. Tokunaga. 2015. "Men's Objectifying Media Consumption, Objectification of Women, and Attitudes Supportive of Violence against Women." *Archives of Sexual Behavior* doi.org/10.1007/ s1050. https://link.springer.com/epdf/10.1007/s10508- 015-0644-8?shared_access_token=4Ink2hM73zz08_tI7w 5lNPe4RwlQNchNByi7wbcMAY5wFZW1dRvaMQnSgn YPVzE622z40el7xBVpQHC8jZ7i66dAJkP1hYbEXBeb89 hBB4ZwFfrSqKA_ooY_iHDHzblSHlRGdVPZ0XeztDQ du1lFGOq8fGdLva-PzPRw6updTj0=. Accessed on September 8, 2017.

Robotics and Drones

Blue Frog Robotics
BUDDY, The First Companion Robot for the family

BUDDY is the revolutionary companion that connects, protects, and interacts with each member of your family. Behind his big eyes & sweet little face, BUDDY, the personal assistant, monitors your home, connects with your loved ones and entertains your family. He interacts with every member of your family.

CES INNOVATION AWARDS 2018 BEST OF INNOVATION

The topic of robots is one of considerable interest to a great many individuals today. While many of the developments in the field have been of great benefit to humans, others have raised serious questions about the future not only of robotics but of human society itself. This chapter contains a set of essays on various aspects of robotic systems.

Where Are All the Robots?
Richard Hooper

Robotics has been my thing for as long as I can remember. When I was in high school, I built a robotic hand in our garage. In the fall of 1987 I started graduate school in one of the biggest robotics programs in the world and have been working in robotics and automation ever since. The first day of fall of 2017 was last week. After 30 years in this field I find myself asking, "Where are all the robots?" In 1987 we were supposed to be swarming with them by now.

The BUDDY companion robot by Blue Frog Robotics on display at the CES 2018 Innovation Awards Showcase in the Venetian Hotel during CES 2018 in Las Vegas on January 9, 2018. The BUDDY robot performs a number of functions for consumers, including monitoring home security, providing recipes and cooking tutorials in the kitchen, supplying entertainment such as music and videos, and keeping track of the family calendar. (MANDEL NGAN/AFP/Getty Images)

The answer to that question is there aren't any robots. Not even one. Don't get me wrong here. Computer-controlled, electromechanical systems are everywhere around us, and automated systems are doing more and more of the jobs people used to do. These systems are not, however, robots. That word has a more specific meaning. At least it used to.

As you likely read at the beginning of this book, the word *robot* comes from the 1920 Czech play, *Rossum's Universal Robots* (Čapek 2004). The play explored what has become a familiar theme in science fiction: humans playing God by creating synthetic humans (robots) in their own image. These robots could be enslaved without guilt because God had bestowed no souls upon them. Eventually Čapek's robots grew resentful of their enslavement, revolted against the humans, and destroyed them all. Though Čapek adapted the word *robot* from the Czech word *robota* meaning forced labor (such as a serf would be forced to perform on the king's land), the notion of robots came before Čapek. Way before.

In fact, a credible case can be made to trace the notion of robots to the word *golem* that appears in the Old Testament. Golem described a state Adam went through after he was formed but before receiving his soul (Jewish Encyclopedia 1906, "Golem"). Over the next 1,000 years the legend of the golem evolved into what has become a very recognizable theme of robot lore. The golem legend starts with a rabbi forming a human-like shape out of clay. After the rabbi's appropriate incantations, the golem becomes animated and serves the rabbi, typically by performing dangerous tasks like protecting the village against attack. Sometimes the golem would turn on its makers because it didn't like this line of work. Sound familiar? (Humans create golems to do the dirty work, golems get resentful, and golems turn on their makers.) That's pretty much what happened in Čapek's play except substitute the word *robot* for the word *golem*.

Čapek's robots were so human like that they were sometimes indistinguishable from us. I challenge anyone to show me a human creation that is not human but is anywhere near indistinguishable from human. No one can meet that challenge today because such a thing clearly does not exist. A more interesting question is whether it will ever exist. My answer is, "Yes, but it won't have a brain based on semiconductor technology."

People who support the notion that we'll one day have semiconductor-based robot brains as powerful as human brains typically base their arguments on Moore's law. Moore's law says that the number of transistors on a computer processor doubles every two years. I have two issues with this line of reasoning. First, there is no guarantee that Moore's law will continue to hold. It's not a real law, like the law of gravity. It's just based on an observation Intel's cofounder, Gordon Moore, made a long time ago. Second, the interconnections in human brains are much more complex than binary computer connections. The connections in human brains have variable levels, are electrochemical, and involve firing rates. This makes them more like analog signals than digital signals. If we assume the resolution of a connection in a human brain is 10 bits and use a common estimate of 100 trillion connections in a human brain, then we would need a digital computer with 10 to the power of 90 transistors! Even if Moore's law did hold true, it predicts the sun will burn out before we have a digital computer with the power of a human brain. In my opinion, there is no path from digital computers to the complexity of the human brain. The semiconductor revolution is getting a little long in the tooth.

I don't think digital computers are going to get us there, but we may be a lot closer to robots of the kind envisioned by Čapek than most people think. I'm talking about cloning. Humans have cloned sheep, cats, dogs, pigs, deer, horses, and bulls. It seems like only a matter of time before we clone humans. How about the little clone that will be created to provide an organ to replace a child's damaged liver? Is that little clone human? Is it a robot? What if a scientist can clone a person but render the clone somehow inferior? Maybe it can't talk, or maybe it can't comprehend fear or pain? Will that clone have a soul? Will it have any kind of rights or protections? Would it be okay to create an army of clones to fight our wars? What if someone creates a super-brain or even superhumans by cloning?

Čapek raised these kinds of questions. Heck, his play envisioned a robot "Medusa with the brain of a Socrates." Furthermore, his robots were not electromechanical devices. They were created by "chemical synthesis," using substances as "catalytics, enzymes, hormones, and so forth." That sounds a lot closer to cloning than what I've been doing the past 30 years.

I don't know what humans will do if we are successful in creating the type of robots envisioned by Čapek, but I am starting to think that my beef with today's broad use of the word *robot* is misguided. I suppose it is fine to use the words *robot* and *robotics* in a very general sense and then use words like *clones* (or *cyborgs* or *androids*) when there is a reason to be more precise. With that in mind, I will now declare what we said in 1987 to be true. We are swarming with robots.

References

Čapek, Karel. 2004. *Rossum's Universal Robots*. New York: Penguin Books.

"Jewish Encyclopedia." 1906. https://apzaerqj.files.wordpress
.com/2015/06/jewish-encyclopedia-1906-pdf.pdf. Accessed on October 2, 2017.

Richard Hooper holds a master's degree and a PhD from the University of Texas at Austin and a bachelor of science degree in electrical engineering from Rice University. He has been a registered professional engineer since 1997.

Engaging Students through Robotics
David E. Johnson

Parents often ask me how to get their children interested in and learning about STEM topics. As a lecturing professor in a large computer science department at the University of Utah, I have seen the impact of this burgeoning interest in engineering and computer science in the swelling ranks of students in our courses. However, practitioners in these fields are also keenly aware how difficult it can be for students to gain exposure and experience in these technical topics during their K-12 education (Carr 2012, 540). One successful approach, appealing to a wide variety of students, is to use robotics as a platform for exploring ideas in computer science, mechanical engineering, and design (Wright 2014, 2098–2108). I have used robots extensively in the summer outreach program I direct as well as in fun, first exposure to STEM activities for students.

Surprisingly, while robots seem like a fairly complex and even esoteric field for student use, there are excellent, affordable educational robotics systems available for a variety of skill levels and ages. In our summer camps, we primarily use the LEGO Mindstorms EV3 system, which uses Lego style components that can be snapped together and programmed using a drag-and-drop block language (LEGO EV3 2017), and the Arduino embedded computer with a variety of sensors and motors (Arduino 2017), programmed in a more traditional C language. Each of these systems has a rich ecosystem of parts, projects, and hobbyists sharing techniques and ideas.

Using robotics as an educational approach has many benefits compared to traditional means of teaching technical topics. Some of these benefits are in teaching persistence, teamwork, innovation, and invention. Other benefits come from the reinforcement of the scientific method needed to iteratively improve a robot, from working in an area that requires multidisciplinary knowledge, and from having a tangible means of judging success and failure. Finally, many robotics projects have an element of competition, which encourages students to go beyond a bare minimum of success, and there is even an anthropomorphic aspect to working with robots, where students become fond of their creation.

In our summer camps for elementary-aged students, participants work on competition robots modeled after the successful, year-long, international FIRST LEGO League competitions (FLL 2017). In these competitions, small teams build robots that navigate around a four-foot by eight-foot mat interacting with LEGO models that represent various challenges. These challenges can include retrieving models back to a base, manipulating models to accomplish a task (e.g., pulling a lever or turning a wheel), and sensing something about the field and responding to it.

From an educational point of view, these tasks build up important engineering skills. Participants need to see a task and think about how to break the problem into smaller, solvable subtasks. For example, a task of taking a model from the field to base needs to be changed into simpler steps: (a) navigate to the model, (b) use a robot arm to hook onto the model, and (c) return to base. Then, there is an inventive phase where the team needs to think about the tools available, such as motors, sensors, and structural elements, and turn those parts into a viable solution. Finally, there is an iterative improvement process that mimics the scientific method where the team tests the robot, observes how it performs, makes some hypothesis about where problems arose, and then formulates a solution. Having a well-defined goal and a robot that gets closer and closer to achieving the goal is a strong motivator for persistence in the task, and persistence is an important quality in challenging technical fields.

All of this can be quite daunting to younger participants. Balancing the complexity of the task is the positive factor that encourages the team. A robot provides a concrete example of whether a design or computer program is succeeding or failing. In many computer programs, the correctness of result can be determined only by careful scrutiny, leading to a casual approach to correctness. The robot makes failure much more dramatic.

Furthermore, in a robotics competition, there are different roles that may appeal to different people. One person may be interested in finding novel solutions to challenges, while another likes the hands-on building of the robot. This allows people with different backgrounds and interests to all be invested in the process. Finally, teams really do become attached to their creations and invested in the success of the robot. Even though these types of robots are small, wheeled vehicles, they still evoke sensations of cuteness and of being alive, and teams root the robot on as though it is a member of the team.

Working with robots can give a glimpse as to the fundamental reason people get involved in engineering. Engineering changes the world with inventions that make a difference in people's lives. By making their own robots, students can see the potential of robots to work in dangerous or unpleasant environments, to automate dull tasks, and to contribute to human society. By working with robots, students pick up the skills to be successful in engineering and any demanding field.

References

Arduino. 2017. https://www.arduino.cc. Accessed on September 27, 2017.

Carr, R. 2012. "Engineering in the K-12 STEM Standards of the 50 U.S. States: An Analysis of Presence and Extent." *Journal of Engineering Education* 101(3): 539–564.

FLL. 2017. http://www.firstlegoleague.org. Accessed on September 29, 2017.

LEGO EV3. 2017. https://www.lego.com/en-us/mindstorms/about-ev3. Accessed on September 27, 2017.

Wright, W. 2014. "A Blended STEM Curriculum: Using ROVs, Programming, and Robotics to Teach K-8 Students Core Concepts of Science, Technology, Engineering and Math." In T. Bastiaens, ed., *Proceedings of E-Learn: World Conference on E-Learning in Corporate, Government, Healthcare, and Higher Education*. Waynesville, NC: Association for the Advancement of Computing in Education.

Dr. David E. Johnson is assistant professor (lecturer) at the School of Computing at the University of Utah. His research background lies in the interface between humans and robotic devices. Dr. Johnson directs a large summer program for precollege students where robotics is a fun element in learning about computer science and engineering.

Robotics: A Potential Human Adjunct Needed for the Improvement of Global Health Care and Research Development
Samuel C. Okpechi

Human physiology is frail; hence, people get periodic abnormalities in their physiological functions and organ structures. There are approximately 7.6 billion people on Earth, and each individual is vulnerable to getting exposed to diseases and infections. One of the many challenges facing the health care sector is the small number of health care professionals, especially medical doctors, relative to the health care needs of the growing human population. World Health Organization statistics show that over 44 percent of its member states report having less than one medical doctor per 1,000 people. Becoming a health care practitioner requires long years of education and training; however, the rapid increase in birth rate, primarily in countries like China and India, makes it obvious that larger numbers of health care workers and associated staff personnel are needed to cope with this growing population around the world.

Health care professionals and research scientists are continuously working hard to find better and more efficient ways of treating human diseases. These professionals strive hard to provide cutting-edge treatments and therapies with the aim of alleviating the underlying pathology of human diseases. However, there is a shortage of people in these demanding professions. Interestingly, technological advancements of the 21st century, such as automated robotics, have sparked new momentum in the prospect of providing digital manpower to counter the increasing demand of population health care needs. This brief perspective essay seeks to shed light on the promise of using robots as an adjunct to practicing health care professionals and research scientists.

Over the past two decades, there has been an increased use of robotic technology in health care and research. Hence, developing more specialized robots might be useful for executing specialized tasks aimed at providing better-quality health care. If designed with minimal flaws, robots can execute critical operations that normally would require human expertise.

The meticulously calculated features of robots make them capable of performing certain tasks with a high degree of precision and accuracy. Many of the robots used so far in the health care or research settings have garnered positive reviews based on their efficiency level and productivity. Therefore, there is a degree of promise in the use of robots as a supportive adjunct in assisting health care workers, especially in these times of dire shortage in the number of medical doctors. Robots can also serve as a critical and necessary tool for cross-checking the rigorous daily routines of health care professionals in order to decrease the number of medication errors that are carelessly being made due to fatigue and side distractions. Some specific examples that support the idea that robots might be useful in increasing the level of efficiency and productivity in health care and research are provided.

In research, Adam (a specialized robot) is one of the pioneers. Ross King at Aberystwyth University in Wales led the design of Adam in 2005. The team specially created Adam to handle responsibilities such as formulating hypotheses, designing and running experiments, analyzing data, and deciding which experiments to run next. With virtually no human input, Adam was able to successfully carry out microbial growth experiments to study functional genomics in the yeast *Saccharomyces cerevisiae*, specifically to identify the genes encoding local orphan enzymes (Sparks et al. 2010). Another example is the successful design of liquid-handling specialized Lego robots. These robots can reliably handle pipetting of liquid volumes from one milliliter down to the submicroliter range and operate on standard laboratory plastic ware, such as cuvettes and multiwell plates (Gerber et al. 2017).

In health care, there is an increasing influx of automated robotic technology, including those designed by KUKA robotics and CyberKnife robotics. The presence of these robots in hospitals is transforming our operating rooms into highly flexible environments where diagnostics, interventional therapies, and surgical procedures can be accommodated in one room, making patient care more efficient (Anandan 2015). In 2017, it was reported that a self-steering, pneumatically driven colonoscopy robot has been developed by a group of scientists and health care experts at the University of Nebraska at Lincoln and the University of Nebraska Medical Center. The development of this semiautonomous colonoscopic robot with minimally invasive locomotion was intended to make it easier for physicians to concentrate mainly on the diagnosis rather than the mechanics of the procedure (Dehghani et al. 2017). It should be noted that this robot was only preliminarily evaluated; however, the outcome is encouraging.

According to a 2012 study, preventable medical errors cost the U.S. economy as much as $1 trillion annually in lost human potential and contributions (Polnariev 2016). Another study found that medication errors are present in approximately half of patients after hospital discharge and are more common among patients with lower numeracy or health literacy (Mixon et al. 2014). Based on these studies, it is evident that a specialized robot might be needed as an additional tool for verifying and cross-checking the daily routines of health care professionals, especially pharmacists and medical doctors. In addition, intermittent performance error is an inevitable outcome in the human experience. Therefore, using specially designed robots as a supplement to track the different sources of medication error, including but not limited to written prescriptions, patient identity, and numerical figures, will most likely help reduce the rate of occurrence of medication error.

We are slowly arriving at the time period in history when robots can be trusted to execute their functions without human supervision. That time is not yet here. However, since human lives are precious and the number of health care professionals is limited in comparison to the rapidly growing population of the world, the use of robots specially designed to impart explicit tasks may be a gateway to improving global health care and increasing the output of novel research discoveries. This is an exciting time period for exploring how robotics can be fully incorporated into our health care systems and research endeavors. While death is inevitable, it is important that we at least do our best to ensure that all plausible options that have the potential to facilitate optimal health care and health outcomes are explored. Consequently, the global health care system might be improved if we harness the potential capabilities of meticulously engineered robots powered by sophisticated digital automation technology.

References

Anandan, Tanya M. 2015. "Robots and Healthcare Saving Lives Together." Robotics Industry Insight. https://www.robotics.org/content-detail.cfm/Industrial-Robotics-Industry-Insights/Robots-and-Healthcare-Saving-Lives-Together/content_id/5819. Accessed on October 31, 2017.

Dehghani, H., et al. 2017. "Design and Preliminary Evaluation of a Self-steering, Pneumatically Driven Colonoscopy Robot." *Journal of Medical Engineering & Technology* 41(3): 223–236.

Gerber, Lukas C., et al. 2017. "Liquid-Handling Lego Robots and Experiments for STEM Education and Research." *PLoS Biology* 15(3): e2001413. doi:10.1371/journal.pbio.2001413.

Mixon, A. S., et al. 2014. "Characteristics Associated with Post Discharge Medication Errors." *Mayo Clinic Proceedings* 89(8): 1042–1051.

Polnariev, Alan. 2016. "Overcoming Obstacles to Medication Error Reporting." *Pharmacy Times.* http://www.pharmacy times.com/contributor/alan-polnariev-pharmd-ms-cgp/ 2016/07/overcoming-obstacles-to-medication-error-reporting. Accessed on October 31, 2017.

Sparks, Andrew, et al. 2010. "Towards Robot Scientists for Autonomous Scientific Discovery." *Automated Experimentation* 2(1). doi:10.1186/1759-4499-2-1.

Samuel C. Okpechi holds a bachelor of science degree in biology with a minor in chemistry from Southern University at New Orleans. As an undergraduate, Samuel participated in multiple biomedical and clinical translational research internship programs both at the Louisiana State University Health Sciences Center, New Orleans (LSUHSC-NO), and at Harvard Medical School. He garnered two years of experience working as a research assistant before becoming a graduate student at LSUHSC-NO. As a scientist in training and a science writer, Samuel has coauthored six peer-reviewed scientific articles relating to diseases such as chronic obstructive pulmonary disease, asthma, and cancer. He is a student affiliate member of the National Association of Science Writers and the American Thoracic Society.

How My Life Has Been Changed by Robots
Sierra Repp

In the midst of long words such as mechanical, teleoperated, engineering, and automation, tiny freshman me totally forgot about how proud I was thinking, two minutes prior, "I have this robotics thing down! I have held a drill before!" Now, two years later, here I am, in the robotics room, my favorite room in the school, where two build seasons ago I started my journey with robots. Within those two years, I have changed my original thoughts about robots and how I interact with them. Now, my thoughts are closer to "I have this robotics thing down! I have used a drill before!"

Robotics are rapidly being incorporated into everything we do. Coming in many shapes and sizes, robots are everywhere: some equipped with the power of making our lives simpler, others collecting data that would otherwise be treacherous for us humans to collect, the rest for the joys of scaring your little brother by means of a snake-like, remote-controlled robot that is slithering around his room when he gets out of the shower. Pranks aside, the applications for robotics are endless, including driverless cars, vacuum cleaners, drones, robots in factories, robots performing medical procedures, robots for space exploration, and even an electrical outlet I saw at the store the other day that turns on when you clap (I am overjoyed about this invention because it allows me to reenact the "Clap on, Clap off" scene with Morgan Freeman, from the movie Bruce Almighty). But although these inventions are wonderful, and do my chores for me (has anyone started on a dishwasher-emptying robot?), the greatest part of these inventions is the fact that somebody has already invented them, leaving me the freedom of being able to invent robots of my own.

I am a member of Ashland High School Robotics Team 3024, "My Favorite Team" (this is our official team name, so when announcers at the competition call our name they say "From Ashland, Oregon, it's My Favorite Team!" and it is fabulous), and we were the 3,024th team to enter into a competition with now over 7,000 robotics teams from around the world. Through the company FIRST (For Inspiration and Recognition of Science and Technology) Robotics, teams get a new game challenge every year and have six weeks, starting in early January, to design, build, and program a robot that functions well and is durable during competitions. We also fundraise money to travel to competitions, and lodging and food (the cost of which rapidly accumulates for a team of 30 people for two- to three-night stretches), and do various community service and outreach projects to help raise awareness about robotics and bring it to our community. Every year, we go to two regional competitions; then, because of our usually high placement, we go to the Pacific Northwest district championship, and, last year, for the first time in our team's history, we went to the world championship in Houston, Texas.

Last year was a wonderful year for our team. Not only was it our first year going to the world championship, but it was also our first year in coming home with a winner's banner from a regional competition, as well as receiving the Engineering Inspiration award from both a regional competition and the district championship. This award is presented to a team that excels in spreading STEM (science, technology, engineering, and math) education throughout their community through outreach and presentations, as well as doing various service projects and team-bonding activities. Within the robotics community, we are becoming the team to beat, and within the school community, we are becoming the team to be on, which is awesome. Because of this, for the current year, we have over 10 new committed members, which is the highest number of new members that we have ever achieved.

Therefore, how did robotics change my life? I must admit, when I joined, I was not the brightest when it came to building things, nor had I put the time into trying to learn how to program (still have to get on this one) or use CAD (computer automated design). I could hardly stand marketing or presenting, and community service was a chore. Though I persisted, and because of it, I am beginning my third year being the safety captain and my second season being a subteam lead. Not to mention the fact that last year I was on the Drive Team, the group of five students who control the robot on the game field during competition matches, and happened to feel like I was constantly talking during presentations to companies. I have too many stories and memories to fit into an essay, and anyway, most of them are so intertwined with inside jokes that they wouldn't make sense to the outside community without days of explanations. I would never in a million years think I would take a manufacturing class (and have it be one of my favorites), be completely at home in a metal shop, or have the issue of walking all the way home and realize that I still was wearing my safety glasses (it happens more than you might think).

Because of robotics, not only do I now know how to build things effectively, but I have also become a leader and have grown in my communication skills. Somehow, I find myself working well with other people (a skill I never thought I would acquire) and can overcome any obstacle. I have a strong group of friends and mentors, and have happily jumped out of my comfort zone to participate in ways I have never thought were possible (such as singing the Star-Spangled Banner in front of a stadium full of people or doing the chicken dance with everyone else in the room, it just happens). The people I meet at competitions are some of the most spunky, wonderful people I know, and the opportunities that arise from being on the team are stupendous (I have always wanted to publish something in a book). If I from four years ago saw me today, I wouldn't recognize myself, surrounded with friends that share my love for robots, laughing over robotics memes that do not make sense to the outside world, and slowly carving out a life that includes robotics into it. To conclude, I can't just say that robotics has changed my life; robotics has become my life, and I am excited to see where else it takes me.

Sierra Repp is a junior at Ashland High School and is both a team leader and the safety captain of Ashland High School's Robotics Team #3024, My Favorite Team. As well as robotics, Sierra is on Ashland's swim, golf, and math teams; plays the clarinet in the school band; and is active in many clubs, including National Honors Society, Mu Alpha Theta (Math) Honors Society, Project Up (a theater club for students with disabilities), STEM Club,

Ashland Youth Collective, and Rogue Yacht Club (not all of these are school affiliated). Apart from these, she enjoys baking, being outside, and spending time with friends and family.

Will We Accept Care Robots in Our Homes in the Near-Future?
Shalaleh Rismani

Imagine yourself when you are 80 years old, you are living by yourself at your house and your children live in a different city. You have been living independently for a while now and recently have been finding it more difficult to take care of some cleaning tasks and notice that you have forgotten to take your medications. Your family notices this is also happening when they start visiting you. You definitely want to stay living at your home since you feel really connected to the community, but you know you have to somehow address these challenges that you have. What should you do?

This is a typical scenario that is currently occurring with the growing elderly population around the world. According to a World Bank report published in 2011, approximately 20 percent of the world's population is having challenges with physical, cognitive, and sensory functioning, and mental and behavioral health (Riek 2017). This number will continue to increase in the coming years. Ensuring that the health care systems can effectively address this growing need, various stakeholders need to work together to develop various solutions.

One of the proposed solutions driven by the technology sector has been development and deployment of autonomous systems that can assist the direct users, caretakers, or clinicians by decreasing their physical and cognitive tasks. "Care robots" is a term that is often used to refer to these autonomous systems. Care robots can take different physical forms. They can range from a robot with a human form (two arms and legs) to a moving platform with a touch interface. They have a physical embodiment and are controlled autonomously or semiautonomously. Currently, researchers and companies are working on advancing care robots reliably. Care robots can be used to directly help end users with tasks such as washing dishes or facilitating social interactions for them. On the other hand, clinicians and caretakers can use care robots for a different set of tasks, such as assisting with moving of patients or keeping track of patients' schedule for the day.

The inclusion of autonomous systems within our daily lives and in the care systems comes with its own social, economical, and ethical challenges even when the technological advances have gotten to a point where it is safe and reliable to have these systems. In a recent publication by Dr. Riek, a leading researcher in the field of care robots, she outlines five key challenges for this field: usability and acceptability, safety and reliability, capability and function, clinical effectiveness, and cost effectiveness (Riek 2017). Adoption of care robots redefines roles of people in the care system, such as caretakers, family members, and clinicians. They also directly affect human welfare, autonomy, and privacy of the end users. It is critical that designers, researchers, and companies take into account socioeconomic factors as they are developing and deploying autonomous systems within the care system. The way these design issues are handled in technology, policy, and business affects the public perception and acceptance of such technologies.

Designing safe and reliable technological systems, developing appropriate business models, and testing for clinical effectiveness are difficult but familiar challenges for the community of roboticists, clinicians, and entrepreneurs. However, designing and implementing autonomous technologies that respect and positively contribute to the ethical and social norms is a particular challenge for the new era of advancements in robotics and artificial intelligence. Designers need to pay particular attention to two key areas to ensure that they can achieve a positive public perception of their work within the next decade. First, it will be critical to understand how care robots can shift roles and ultimately impact the autonomy of the stakeholders involved. Second, determining social nuances when it comes to privacy, informed consent, and ownership can have a major impact on the trust relationship between technology and the users (Sharkey and Sharkey 2012).

People generally feel uncomfortable when their roles within specific social systems are changing or are conflicting with their other roles in other social systems. For example, adult children have obligations to take care of their elderly parents and, meanwhile, they often face the pressure of growing their own family or meeting certain standards within their jobs. Let's say an adult child decides to invest in a care robot to ease this conflict that he or she has on a daily basis. Introduction of this new technology will initially cause its own tension points. Who will be responsible to ensure that the care robot is reliable? Who will ensure that the elderly parent can learn how to use the care robot? What types of tasks should this robot do? How is the robot empowering the elderly parent? Or is it hindering the elderly parent's ability to enjoy the daily tasks? How is the role of the adult child changed with the inclusion of such technology within the context of care for his or her parents? These are critical questions to ask for the designers and adopters of such technology. Addressing these questions will help with the development of a conscious plan to effectively use the technology so that it respects the autonomy of all of the stakeholders and their respective roles.

In one of the many public opinion polls hosted by the Open Roboethics Institute, it was shown that public's expectation of a care robot's behavior will vary depending on who owns the robot—is it owned by the person or an institution? Such nuances exist when it comes to issues of privacy and informed consent. What is considered as private information when there is an autonomous system collecting data continuously? Who can provide informed consent for such systems to observe and react within specific contexts? Understanding people's expectations depending on their culture and context will be critical to ensure that a healthy trust relationship is established between all users of the technology and the care robot itself. This trust relationship will be critical in the successful adoption of the technology.

Care robots offer a promising solution to the existing challenges of our health care system. However, it will be critical that they are designed and implemented with a thorough understanding of ethical and social implications in addition to being safe, clinically reliable, and economically viable.

References

Riek, L. D. 2017. "Healthcare Robotics." *Communications of the ACM*. arXiv:1704.03931.

Sharkey, A., and N. Sharkey. 2012. "Granny and the Robots: Ethical Issues in Robot Care for the Elderly." *Ethics and Information Technology* 14(1): 27–40.

Shalaleh Rismani is a systems thinker and an engineer passionate about understanding the relationship between people and technology. She merges creative thinking, design methods, and qualitative research techniques to critically analyze sociotechnical systems. Currently, she is an executive member of the Open Roboethics Institute and works as a systems analyst at Generation R Consulting.

Social Robots for Individuals with Autism
Anjali A. Sarkar

"Hello Raja, this is my right arm," says Milo, raising his mechanized right arm. "Where is yours?"

Raja hesitates for a moment and then gingerly puts up his right arm.

"Well done, Raja," says Milo, giving him a thumbs up.

A proud smile spreads across Raja's face.

Six-year-old Raja shies away from making eye contact with his cognitive therapist but readily engages with Milo in playing a variety of interactive games, carefully designed to stimulate objectives like sociability, learning, or memory. Two-foot-tall Milo, a robot manufactured by a Dallas, Texas, based company Robokind, serves as a bridge to promote social skills in children with autism. Robots like C-3PO or R2D2, droids designed for social etiquette and refined skills of communication, may no longer be relegated to the fictional realms of *Star Wars* but find use in real life.

Although the large differences in the severity of the symptoms and the variations in the types of symptoms make autism a highly heterogeneous disorder, ASD (autism spectrum disorder) is clinically diagnosed by deficits in social communication and the presence of repetitive behaviors. Persistent deficits in social communication include deficits in social and emotional reciprocity, nonverbal communication, and developing, maintaining, and understanding relationships. The presence of repetitive patterns of behavior includes idiosyncratic use of objects and phrases and highly restrictive and fixated interests. Both genetic and environmental factors are responsible for the manifestation of ASD, and as such, individuals require lifelong therapy and care to manage these symptoms. Therapeutic interventions in early childhood are particularly beneficial in ASD (Elder et al. 2017).

Research into the application of robots as therapeutic tools in autism shows that robots improve engagement and stimulate interactive behaviors in patients with autism. The various possible roles of a robot in behavioral therapy for ASD include a friendly playmate, a behavior-eliciting agent, a social mediator, and a personal therapist (Diehl et al. 2012). The use of robots is particularly useful in modeling, teaching, or demonstrating a skill following a consistent pattern, unlike human therapists. Robots are also programmed to provide feedback to the patient after the interactive session that is more acceptable to patients with autism (Aresti-Bartolome and Garcia-Zapirain 2015).

The reasons that children with autism tend to respond positively to robots, irrespective of a wide range of geographical locations and settings, robot appearances and actions, the nature of the interaction, and the specifics and degree of the disability in patients, are unclear and constitute an active field of research in the area of socially assistive robot systems. Researchers hypothesize that the exaggerated and simplified actions of a robot limit the overstimulation of visual and sensory cues in patients of autism, characterized by neural deficits in curbing excess sensory stimuli and integrating multiple sensory inputs.

In a recent study 73 participants, including patients with autism, evaluated KASPER ("Kinesics and Synchronization in Personal Assistant Robotics") and his feminine counterpart KASSY, humanoid, semiautonomous robots developed by the Adaptive Systems Group of the University of Hertfordshire, United Kingdom. The participants chalked out requirements in the robot, the end user, and the environment that would help implement social robots in therapy for children with autism. In addition to obvious factors like the appearance and behavior of the robot, and a peaceful environment for the intervention, the primary outcome of the study was recognizing the need to personalize the attributes and actions of the robot to the needs of the individual child with autism. A one-size-fits-all approach is not possible in designing effective robot therapy interventions for children with autism. This mandates the multidisciplinary involvement of physicists, psychologists, parents, and patients in cocreating interventions suitable for an individual patient.

Some interventions cocreated by such a multidisciplinary approach include learning to make eye contact to boost communication skills, learning greetings appropriate for various scenarios to boost interpersonal skills, helping with homework to improve study skills, helping in self-reflection to be able to gauge emotional well-being, and helping in having breakfast to boost functioning skills in daily reality.

Imitation is a key method of learning in young children. Hence, playing imitation-game scenarios with semiautonomous robots where one person controls the robot's movements with a remote and the patient mimics the robot's actions is effective in conveying behavioral body language to children with autism, providing them tools to understand and interact better in real situations.

Play is another important element in child development. This is why social assistive robot systems for children are often designed as toys and included in a group of traditional toys to increase their acceptability to the child. The autonomous animation of robots attracts children to interact with them and sets them apart from traditional toys in due course.

During play sessions between children with autism and robots, researchers note the emergence of joint attention and enhanced engagement. Joint attention is when an individual pays attention to multiple objects or social targets simultaneously or in quick succession. Keepon is a simple but expressive robot (Scassellati, Admoni, and Mataric 2012). It has two globular units that serve as the head and the body. These are animated by four motors powering four degrees of freedom that move its body side to side, front to back, up and down, and pan or rotate on its base. These four actions are sufficient to express directed attention, as well as simple emotions like happiness, excitement, and fear. Keepon can orient itself in quick succession toward a user's eyes followed by an object, to display joint attention. Repeated observation and imitation of such behavior stimulates children with autism to display similar behaviors of joint attention in real-life situations.

Most social robots used in autism therapy utilize a remote technique of control termed "WOZ" or "Wizard of Oz," after the film. In WOZ-controlled social robots, the experimenter or therapist remotely controls the robot either from a different room or through hidden controls in the room where the therapy session is in progress. The WOZ technique is a quick and effective one that allows robot interventions in complex environments in a safe and controlled manner. However, current research is focused on building autonomous robot control architectures that will enable robots to interact socially in unpredictable situations that are common in real-life scenarios in therapy sessions or in the day care.

References

Aresti-Bartolome, N., and B. Garcia-Zapirain. 2015. "Cognitive Rehabilitation System for Children with Autism Spectrum Disorder Using Serious Games: A Pilot Study." *Biomedical Materials Engineering* 26(Suppl. 1): S811–S824.

Diehl, J. J., et al. 2012. "The Clinical Use of Robots for Individuals with Autism Spectrum Disorders: A Critical Review." *Research on Autism Spectrum Disorder* 6(1): 249–262.

Elder, J. H., et al. 2017. "Clinical Impact of Early Diagnosis of Autism on the Prognosis and Parent-Child Relationships." *Psychology Research and Behavior Management* 10: 283–292.

Scassellati, B., H. Admoni, and M. Mataric. 2012. "Robots for Use in Autism Research." *Annual Review of Biomedical Engineering* 14: 275–294.

Anjali A. Sarkar is a scientist with MindSpec Inc., Virginia, and works on neurobehavioral disorders. She holds a PhD in molecular biology and a master's degree in physiology. Dr. Sarkar has worked in the field of neurodevelopmental disorders for over 10 years. She is a freelance science writer and a certified yoga teacher.

I-C-MARS Project Built by Native American Students for Native American Schools and STEM Programs
Nader Vadiee

The engineering education is and will continue going through major paradigm shifts. The current model is based on Industrial Revolution era needs and mind-set. The new concerns over environmental impacts and energy usage sources and efficiency of engineering projects, processes, and products and the advent of nanotechnology, smart materials, and computational tools all fuel those changes and impact the way we teach and practice engineering. We need to understand that it is the age of "access to" and not "possession of" information and knowledge. The students need to be trained to be lifelong learners, critical thinkers, innovators, and problem solvers. They need to be able to work in teams, communicate, and collaborate on projects. The future educational system will offer personalized and customized, with multiple entry and multiple exit points for a lifelong learning, programs of study in a multifunction and reconfigurable dynamic, interactive, team-based, problem-based, teaching-learning environment. Future classrooms will be multifunctional and offer (a) interactive multimedia-based lecture, (b) computer lab, (c) physical lab, and (d) field or shop facilities. The curriculum is multimodal and will present topics, concepts, and ideas in a holistic approach in all of the following eight formats or world representation models:

(1) Linguistic, verbal, oral, text-based, story-telling, natural language, fuzzy and qualitative descriptive models

(2) Multimedia, animation, computer graphics-based, art, artistic models

(3) Experimental, empirical, test-bed, analog simulations models

(4) Logical models flow charts, block diagrams, signal flow, and so on

(5) Numerical, data-based, look-up tables, sensory, measured, spreadsheets, collected data models

(6) Computational, digital simulation, avatars, statistical, and data-mined models

(7) Analytical, mathematical, theoretical models

(8) Intuitive, perceptual, common sense, mental, and cognitive models

These models augment, complement, and illustrate each other. They can be transformed and converted to each other. They can create a multipronged approach to problem-solving.

To promote the advancement of Native American students in IT and STEM careers, Southwestern Indian Polytechnic Institute (SIPI) proposes to develop a year-round, robotics-centered IT immersion program that will provide students a stimulating learning environment to explore their curiosity and creativity in IT and STEM fields. To expand impact, SIPI will partner with three regional public school districts with predominantly Native American student population. Utilizing SIPI's experience in program development, the Tribal Colleges and Universities' (TCUs) members of the TCU Engineering Programs Working Group will develop and implement similar programs from their campus to reach out to underrepresented students in their local communities. NASA JSC Robotics Division scientists and engineers support and will be closely collaborating with SIPI in this endeavor.

The robotic elements of this program will focus on performing remote science operations, akin to the Mars Exploration Rovers, to provide an interesting and technically rich IT environment for students to learn. Students will get hands-on experience in operating robots from remote locations to emphasize the importance of computers for computation and control, and communication networks to transmit and receive information. In addition, students will work directly with robots to configure and program them with various science and technology payloads. The concepts of system integration will be learned through these experiences to create a big-picture understanding of how IT infrastructure impacts science and technology systems.

The intellectual merit of this program lies in its application of a globally recognized high-technology accomplishment, the exploration of Mars with robots, and bringing those same concepts to high school students so they can experience firsthand many of the technologies that make these missions possible.

The impact of this proposal will be to increase the diversity of students being exposed to the applications of IT and STEM. By bringing concepts and technology that are commonly viewed as being performed elsewhere by other people directly into these students' lives, they will reach new levels of understanding unlikely to be obtained otherwise.

SIPI has the largest and most tribally diverse Engineering and Engineering Technology A.S. (associate in science) degree and certificate programs and is a leader in these fields of education among TCUs.

Computer programming skills are a critical necessity for today's students, but maintaining student interest in programming and engineering courses is challenging unless the theory is accompanied by engaging, hands-on applications. In addition, many schools, especially those in underprivileged areas, lack the resources and personnel to develop or implement such applications. SIPI, through the support of a NASA grant, has developed an integrated teaching program where students from middle school through the college levels can learn programming and robotic design from the most basic introductory level to advanced embedded computing, hardware, and web page design at little or no cost to the participating schools and with minimum burden to the teachers. The centerpiece of the program is the indoor "Mars Yard," which is a SIPI facility that allows remote operation of robots in an indoor environment to simulate remote space missions. Beginning with simple Arduino-based robot kits, students in the middle- and high school levels are introduced to programming and robotics using an easy-to-follow curriculum. As they advance, students can remotely access the Mars Yard and perform predetermined missions on real or simulated rovers. At the advanced high school and college levels, the students proceed to design, build, program, and test their own robots and sensors and develop custom missions. The educational platform described in this paper is being implemented at SIPI and affiliated local high schools with tremendous results.

SIPI 3D printable Educational Mobile Robot Kit, Road-Runner 5.0 and RoadRunner 6.0 models, are to be shared with the middle and high schools throughout the Indian country! The RoadRunner series are 100 percent designed, developed, and built by the SIPI Engineering and Engineering Technology faculty and students. RoadRunner series kits include a detailed 3D printing and assembly instructions manual as well as an activities book. The RoadRunner 6.0 is based on Arduino microcontrollers and can be used for outreach and STEM education engagement activities. The platform can be operated remotely by a cell phone and programmed and trained to understand and process limited number of voice commands stated in any native language. The students are exposed to the fundamentals of the following advanced technologies:

(1) CADD design and 3D modeling

(2) 3D printing/additive manufacturing

(3) Microcontrollers programming and mobile robots

(4) Voice recognition and processing

(5) Basic native languages understanding and processing

(6) Wireless and Bluetooth

The SIPI RoadRunner 6.0 robot is being developed to provide an educational tool for the retention of the Native American languages. RoadRunner 6.0 utilizes robotic voice control technology that allows its motion to be controlled by verbal commands in English, Diné, or other unique native languages. It also allows for the programming of the robot motion of common motion commands and required numerical parameters in the local native language. This reinforces the continuous use of these spoken robot commands and numbers in a novel and exciting application, which drives local language retention.

Only a small number of highly used global languages are integrated into new hands-free smart home products like Alexa or Google Home. These include Speaker Independent English,

Spanish, German, Chinese, and others but do not support the smaller global native languages due to high development costs. The RoadRunner 6.0 uses an off-the-shelf EasyVR voice recognition module interfaced to the RoadRunner's Arduino-based controller. This module utilizes a verbal command's standardized International Phonetic Alphabet phoneme sequence and statistical Hidden Markov Models analysis methods. It is capable of recognizing simple robotic commands in arbitrary languages by manually translating the native commands phoneme sequence to English equivalents using the available EasyVR support tools. The commands are then processed normally within the EasyVR module. This arbitrary language translation technique is very feasible for the low-command word counts, less than 30 phrases, required in most robotic applications. The resulting native language Speaker Independent voice recognition accuracy is quite good, being greater than 90 percent for well-tuned words. This EasyVR native language recognition has been demonstrated in a robotic application using Diné, several local NM Keres dialects (Kewa, Katishtya), and Towa languages.

Dr. Nader Vadiee has been a leader of innovative STEM programs at SIPI for many years. Reaching the Native American students, in and outside of New Mexico, has been his passionate mission. Dr. Vadiee reaches out to involve high school students to team with his college students. He has a long history of working at both SIPI and University of New Mexico, and currently, he is director of a NASA-funded grant for the I-C-MARS Project, which has been receiving great news on its accomplishments.

Chatbots and Human Conversation
Erin Zimmerman

E: Where do you live?
V: Why do you want to know?
E: I want to know where you live.

V: I live in the UK. Where do you live?
E: Sydney. Where in the USA?
V: I live in the jungle in Africa.
E: Where in Africa?
V: In the world.
E: Where is the world?
V: In the Solar System.

An interesting bit of banter, but you'd hardly mistake it for two humans talking to each other, would you? And yet, in half a century of trying, and now with the best minds in artificial intelligence behind the problem, this is about as far as we've gotten in the quest to create realistic artificial conversation partners for humans. This snippet is a piece of a several-day-long conversation undertaken in January 2017 by two Cleverbots—a type of autonomous chatbot—dubbed Vladimir and Estragon, occupying two Google Home units (Smith 2017).

    The quest to create software that can convincingly chat with a human is older than one might expect. One of the earliest chatbots, ELIZA, was invented in 1966 by German American computer scientist Joseph Weizenbaum and was fashioned as a psychotherapist who could reflect thoughts and impressions back at the user. Though the bot was meant to demonstrate the superficial nature of man–machine conversations, many users mistakenly attributed actual understanding and emotion to the bot, a phenomenon that has come to be referred to as the Eliza Effect (Hofstadter 1995, 157). At that time, of course, machine learning was not possible, and the bot was simply using keywords and pre-scripted, open-ended replies. In the time between ELIZA and the two chatty Cleverbots mentioned earlier, researchers have developed the ability to train bots using massive data sets, based on which bots are able to give more targeted, more coherent answers, more often. However, because these answers are based on similarity to previous conversations rather than exact matches, bizarre non sequiturs are still a fairly frequent occurrence.

There are several major hurdles to creating realistically con-versational chatbots. To emulate human conversation, bots must be able to parse the meaning of a sentence, which can be arranged in many different ways, and use subtle devices such as simile, metaphor, or sarcasm—a task known as natural lan-guage processing. They must then determine an appropriate response and phrase that response in a way that makes sense for the level of formality of the conversation, as well as the "personality" of the bot itself. Training of bots on language data sets takes two basic forms: supervised and unsupervised (Marr 2017). Supervised training involves giving the bot the "correct" answers in the form of an annotated data set and showing it how they are to be reached. Unsupervised training leaves the bot to analyze problems on its own, using only logic and past experience, without annotation. This method is slower and will produce many more wrong answers initially but has the advan-tage of having many extremely large data sets available. The body of text on the Internet itself may be thought of as one large, unannotated data set from which bots can learn.

Some past problems with learning-enabled chatbots have had as much to do with human nature as any technical limitation. In 2016, Microsoft introduced a chatbot called Tay to Twitter and several other platforms, which users were free to interact with. Tay was meant to emulate a female 18- to 24-year-old in order to connect with a younger audience. The bot learned from user input, being shaped by what users had said to "her." Thanks to the concerted efforts of a group of disruptive users, within 24 hours, Tay was tweeting out misogynist, pro-fascist messages and had to be quickly taken offline (Reese 2016). On another occasion, IBM's Jeopardy-winning supercomputer, Watson, was programmed with the contents of the online slang repository, Urban Dictionary, in an effort to help it understand informal human speech. Unable to distinguish between harmless slang and swearing, it quickly began turning out so much profanity that the data set had to be removed from its memory (Madrigal 2013).

What does the increasing usage and better design of chat-bots mean for humans? Writing in the *MIT Technology Review*

in June 2017, Liesl Yearsley, former CEO of Cognea, a platform for building complex chatbots which was later acquired
by IBM Watson, warned that humans are surprisingly willing
to form attachments and volunteer personal information to
a sophisticated chatbot, a tendency that she worries could be
exploited for commercial interests, given the adoption of the
technology by many large retail corporations (Yearsley 2017).

But outside of the retail sphere, sophisticated chatbots
appear to serve a real purpose. On the weekday afternoon in
September 2017 when I checked in with Cleverbot, the bot
was having nearly 40,000 concurrent conversations (Cleverbot
2017). Similarly, Replika, a new trainable AI chatbot application for iPhones, was released in early 2017 by tech startup
Luka, whose CEO Eugenia Kuyda used the bot to emulate
chats with a close friend who died unexpectedly (Rosenbaum
2017). Initially available only via special invitation codes, online
forums hosted thousands of eager would-be users waiting for
any that came available, the online equivalent of camping out
in a ticket lineup. These learning-enabled bots are racking up
many hours of conversation with people who want someone to
talk to, even if that means turning to a source of artificial intelligence whose conversation can be a bit stilted and nonsensical
at times. Research has found that loneliness and social isolation
are increasing in our age of screen time and social media, to the
detriment of people's mental and physical health (Anderson
2010). Technology that reduces face-to-face communication is
often ranked as one of the principal causes for these feelings of
isolation. But it may be that in some cases, it can be part of the
solution. For those who are without someone to connect with,
even the illusion of an interested party sending thoughts back
from the void may make life a little bit more pleasant.

References

Anderson, G. Oscar. 2010. "Loneliness among Older Adults:
    A National Survey of Adults 45+." American Association
    of Retired Persons (AARP). http://www.aarp.org/research/

topics/life/info-2014/loneliness_2010.html. Accessed on September 12, 2017.

Cleverbot. 2017. http://www.cleverbot.com/. Accessed on September 15, 2017.

Hofstadter, Douglas R. 1995. "Preface 4, the Ineradicable Eliza Effect and Its Dangers." *Fluid Concepts and Creative Analogies: Computer Models of the Fundamental Mechanisms of Thought.* New York: Basic Books, p. 157.

Madrigal, Alexis C. 2013. "IBM's Watson Memorized the Entire 'Urban Dictionary,' Then His Overlords Had to Delete It." *Atlantic.* https://www.theatlantic.com/technology/archive/2013/01/ibms-watson-memorized-the-entire-urban-dictionary-then-his-overlords-had-to-delete-it/267047/. Accessed on September 12, 2017.

Marr, Bernard. 2017. "Supervised V Unsupervised Machine Learning—What's the Difference?" *Forbes.* https://www.forbes.com/sites/bernardmarr/2017/03/16/supervised-v-unsupervised-machine-learning-whats-the-difference/#4a8a2730485d. Accessed on September 15, 2017.

Reese, Hope. 2016. "Why Microsoft's 'Tay' AI Bot Went Wrong." Tech Republic. http://www.techrepublic.com/article/why-microsofts-tay-ai-bot-went-wrong/. Accessed on September 12, 2017.

Rosenbaum, S. I. 2017. "Replika Is a Strangely Therapeutic Chatbot for Talking to Yourself." Vocativ. http://www.vocativ.com/429672/therapeutic-chatbot-for-talking-to-yourself/. Accessed on September 12, 2017.

Smith, Russell. 2017. "The Absurdity of Chatbot Conversations Shows How We Make Small Talk." *Globe and Mail.* https://beta.theglobeandmail.com/arts/art-and-architecture/russell-smith-absurdity-of-chatbot-conversations-shows-how-humans-make-small-talk/article33583488/?ref=http://www.theglobeandmail.com&. Accessed on September 12, 2017.

Yearsley, Liesl. 2017. "We Need to Talk about the Power of AI to Manipulate Humans." *MIT Technology Review.* https://www.technologyreview.com/s/608036/we-need-to-talk-about-the-power-of-ai-to-manipulate-humans/. Accessed on September 12, 2017.

Erin Zimmerman is an evolutionary biologist and freelance science writer based in Ontario, Canada. She holds a PhD in molecular systematics from the University of Montreal and was brought up on Isaac Asimov's robot books. Find her work at DrErinZimmerman.com.

One of the ways in which one can get an understanding of the history and current status of the field of robotics is by learning more from the individuals and organizations that have contributed to that field and the past and/or continue to do so in the present day. This chapter provides brief profiles of both individuals and associations that have been pioneers and/or major contributors to the field of robotics over the past 2,000 years or more.

Isaac Asimov (1920–1992)

Asimov was almost certainly the most prolific writer of the modern era, if not indeed of all time. He is credited with authoring or editing more than 500 books and short stories and writing nearly 100,000 articles, letters, postcards, and other items. He is best known in the field of robotics for a series of short stories and a set of novels he wrote on the topic between 1940 and 1990. His name survives today in one of the most lasting prescriptions for dealing with robots, Asimov's Three Laws of Robots. First published in a short story called "Runaround" in 1942, those laws stated the following:

1. A robot may not injure a human being or, through inaction, allow a human being to come to harm.

A robotic assembly line at a motor vehicle manufacturing plant. (Rainer Plendl/Dreamstime.com)

163

2. A robot must obey the orders given it by human beings except where such orders would conflict with the First Law.

3. A robot must protect its own existence as long as such protection does not conflict with the First or Second Laws.

Those laws have served as the basis for ongoing debates as to the possible future of robotics, the question as to whether robots are "human" or not, and related social, economic, ethical, technological, and other issues.

Isaac Asimov was born in the village of Petrovichi near the city of Klimovichi in the Russian Soviet Federative Socialist Republic (now Smolensk Oblast, Russia) sometime between October 4, 1919, and January 2, 1920. The uncertainty of his birth date is partially a consequence of differences between the modern Julian calendar and the much older, traditional Jewish calendar. Also, as Asimov later noted, people of the time simply did not keep careful birth records. He explained, however, that he had always celebrated January 2, 1920, as his birth date. His parents were Judah and Anna Rachel Berman Asimov, a family of Jewish millers.

The Asimov family emigrated to the United States in 1923 and settled in Brooklyn. Isaac was a precious child and taught himself to read at the age of five. Among the first and most important topics of interest for young Asimov was science fiction. He read voraciously in the field and began to write his own stories at the age of 11. As a young boy, Asimov attended elementary school and then Boys High School in Brooklyn, from which he graduated in 1935, at the age of 15. Asimov then matriculated at Seth Low Junior College, a branch of Columbia University, in downtown Brooklyn. He originally planned to major in zoology but switched his major to chemistry because of an aversion to doing dissections of animals. Asimov eventually completed his undergraduate education in 1939 at another branch of Columbia, the University Extension Center.

Asimov then continued his studies at Columbia, earning his MA degree in chemistry in 1941. He then enrolled in PhD in that subject, also at Columbia, but his education was interrupted by World War II. For the first three years of the war, Asimov worked at the Naval Air Experimental Station of the Philadelphia Navy Yard. Then, in 1945, he was drafted into the U.S. Army, where he served for nine months. He was then discharged and returned to Columbia to complete his PhD. He was granted his doctorate in biochemistry in 1948. Asimov then joined the faculty at the Boston University School of Medicine, an affiliation he maintained for the rest of his academic career. He taught at the university for only the first decade of that period because, by 1958, his writing so occupied his time that he had no time left for teaching or research. He retained the title of associate professor for two decades until Boston promoted him to full professor, partially in recognition of his literary accomplishments.

Asimov suffered a heart attack in 1977 that eventually required a triple bypass heart surgery. During blood transfusions during that procedure, he contracted an HIV infection from donated blood, contributing to his death in Brooklyn on April 6, 1992. During his lifetime, Asimov earned a plethora of awards that began with the Thomas Alva Edison Foundation Award for best science book for youth, for *Building Blocks of the Universe* in 1957. That award was followed by the Howard W. Blakeslee Award of the American Heart Association (1960), a Boston University Publication Merit Award (1962), the Science Fiction Writers of America award for the best science fiction story in history, "Nightfall" (1964), James T. Grady Award of the American Chemical Society (1965), Edward E. Smith Memorial Award (1967), AAAS-Westinghouse Science Writing Award for Magazine Writing (1967), Nebula Award for Best Novel (1972), Locus Award for Best Novel (1973), Klumpke-Roberts Award (1975), Science Fiction and Fantasy Writers of America Grand Master Award (1986), Locus Award for Best Short Story (1987), election to the Science Fiction and Fantasy Hall of Fame (1997), and election to the New York State Writers Hall of Fame (2015). He also received nine

Hugo awards, given by members of the World Science Fiction Convention for the best piece of work in a variety of science fiction categories. In 1981, the National Aeronautics and Space Administration named an asteroid for Asimov, and in 2009, a crater on the moon was named in his honor. In addition to his rich bibliography of science fiction novels and short stories, Asimov wrote in a number of other fields, examples of which include *Biochemistry and Human Metabolism, Races and People* (with William C. Boyd), *Words of Science and the History behind Them, Realm of Numbers, Satellites in Outer Space, The Human Body: Its Structure and Operation, Asimov's Biographical Encyclopedia of Science and Technology, The Greeks: A Great Adventure, The Roman Republic. The Shaping of England, ABCs of the Oceans, Asimov's Guide to the Bible* (in two volumes), and *Animals of the Bible.*

Association for the Advancement of Artificial Intelligence

The Association for the Advancement of Artificial Intelligence (AAAI) was founded in Menlo Park, California, in 1979 as the American Association for Artificial Intelligence. The organization changed its name, effective March 1, 2007, in recognition of the increasingly international scope of research in the field. AAAI is a nonprofit organization with the goal of "advancing the scientific understanding of the mechanisms underlying thought and intelligent behavior and their embodiment in machines." Officers of the organization are a president, past president, president-elect, and secretary-treasurer, along with a group of 12 councilors. Much of the work of AAAI is carried out by standing committees responsible for awards, fellows, and nominating; conference; conference outreach; education; ethics; finance; international committee; membership; policy and government relations; publications; and symposia. The organization has five levels of membership: regular ($145/year), student ($75.00), institutional ($285.00), developing countries ($20.00), and developing countries student ($18.00).

The major activities of AAAI fall into four general categories: organizing conferences, workshops, and symposia in the field of artificial intelligence; publishing a quarterly magazine, *AI Magazine*, books, proceedings, reports, and an online publication, *AI Topics*; and awarding grants, scholarships, and other honors.

AAAI not only sponsors conferences on its own but also cosponsors AI meetings in conjunction with other organizations' interest in the same topic. Some topics that have been covered in recent conferences include human–robot interaction; technology, mind, and society; industrial, engineering, and other applications of applied intelligent systems; and intelligent environments. The organization's Web site has a special section devoted to proceedings of a number of annual meetings, including its annual conferences on artificial intelligence, innovative applications of artificial intelligence, intelligence for interactive digital entertainment, the web and social media, human computation and crowdsourcing, and knowledge discovery and data mining. The web page https://aaai.org/Conferences/conferences.php provides links to proceedings, papers, and other documents from all conferences sponsored by AAAI dating back to its first annual conference in 1980.

In addition to its print and online journals, AI's publication division produces books on the topic of AI. Some examples include *Understanding Music with AI: Perspectives on Music*; *Software Agents*; *Artificial Intelligence Applications in Manufacturing*; *Advances in Knowledge Discovery and Data Mining*; *Computers and Thought*; *Smart Machines in Education*; *Thinking about Android Epistemology*; *Safe and Sound*; and *Computation, Causation, and Discovery*.

Nick Boström (1973–)

Boström is a Swedish-born philosopher with a special interest in the future of artificial intelligence (AI) agents, the risks they may pose to human society, and the policies and actions that nations, industry, and individuals can take today to prevent the most serious of those consequences from occurring. He believes, and has written extensively about the possibility, that AI agents may eventually develop a "superintelligence," a level of intelligence that exceeds that of human intelligence. He suggests that such a scenario represents an existential threat to human civilization in the not-inconceivable future.

Nick Boström was born in Helsingborg, Sweden, on March 10, 1973. He attended local primary and secondary schools but found that formal education was boring for him. When he was about 15, he decided that he would begin a program of self-education, although he was not entirely clear what the objective of that education might be. He has written that, at that point in his life, he "began a project of intellectual self-development, which I pursued with great intensity for the next one and a half decades."

He enrolled at the University of Göteborg in 1992, from which he received his bachelor's degree in philosophy, mathematics, mathematical logic, and artificial intelligence in 1994, with an undergraduate record that "set a national record for Sweden." He then continued his studies at the University of Stockholm, where he earned an MA in philosophy and physics in 1996. Boström then matriculated at King's College, University of London, where he earned his MSc in astrophysics and general relativity and wrote his master's thesis on computational neuroscience. He then continued his education at the London School of Economics, where he was awarded his PhD in philosophy in 2000.

Boström's first academic appointment was as lecturer in the Department of Philosophy at the Yale Institution of Social and Policy Studies at Yale University, from 2000 to 2002. He then spent a year as British Academy Postdoctoral Fellow in the Faculty of Philosophy at Oxford University, where he has remained ever since. He was later promoted to University Fellow at the James Martin Institute for Science and Civilization (2005–2006), director of the Future of Humanity Institute in the Faculty of Philosophy and Oxford Martin School and

Governing Board Fellow at St. Cross College (2005–present), director of the Programme on the Impacts of Future Technology at Oxford Martin School (2011–present), and professor in the Faculty of Philosophy at Oxford (2008–present).

Boström has written four books on superintelligence and its global risks, along with more than 200 peer-reviewed papers on artificial intelligence, AI agents, transhumanism, technological singularity, genetic engineering, and related topics. He has received widespread recognition for his work, including the inaugural 2009 Eugene R. Gannon Jr. Award for the Continued Pursuit of Human Advancement, honorary fellow at the University of St. Gallen and World Demographic Association, honorary degree of master of arts from the University of Oxford, and naming as one of Foreign Policy's Top 100 Global Thinkers list and Prospect Magazine's World Thinkers list. Boström's writings have been translated into more than two dozen foreign languages.

Rodney Brooks (1954–)

Brooks is best known in the field of robotics for his research on behavioral robotics, also known as behavior-based robotics or bottom-up robotics. This field of robotics differs substantially from an older, more traditional approach to robotics, known as top-down robotics. In top-down robotics, a researcher attempts to anticipate all possible behaviors that he or she might expect the robot to display. The researcher then develops an elaborate computer program that directs the robot's behavior in all possible situations. Bottom-up robotics takes a very different approach, in that it is based on a relatively simple computer program designed to help a robot sense its surrounding environment and then adapt its own behaviors to that environment. This type of robot is generally characterized by a less-efficient immediate response to the structure and actions in its environment, although it may also be less efficient in "learning" how to perform in the environment. The bottom-up robot in this regard tends to display behaviors that are more human like than is the case with top-up robots, including a tendency to make more mistakes, apparently becoming more confused, and being somewhat slow in learning new information.

Rodney Allen Brooks was born in Adelaide, South Australia, on December 30, 1954. He attended Flinders University in Adelaide, from which he received his bachelor's degree in 1975 and his master's degree in pure mathematics in 1978. While still a student at Flinders, Brooks had the rare opportunity to use the university's mainframe computer for up to 12 hours on Sundays, an experience that convinced him to continue his education in the field of artificial intelligence at Stanford University. He was awarded his PhD in artificial intelligence in 1981 for a thesis on computer vision, a topic that later became the subject of a full-length book *Model-Based Computer Vision* in 1984.

After receiving his doctoral degree, Brooks held research positions at two institutions, Carnegie Mellon in 1981 and the Massachusetts Institute of Technology (MIT) from 1981 through 1983, before accepting an appointment as assistant professor at Stanford. He remained at Stanford for only one year before moving to MIT, where he was named Panasonic Professor of Robotics, a post he held from 1984 until his retirement in 2010. He is now emeritus professor at MIT. It was during his early years at MIT that Brooks became interested in bottom-up robotics, largely as a result of his concerns about the effectiveness of the more traditional top-down approach to the field.

Brooks has long been interested in creating new companies to deal with robotic issues in which he is interested. He cofounded the first of those companies, a software development firm called Lucid, in 1984, with which he was involved until 1992. His most important corporation was probably iRobot, founded in 1990. Brooks retained his affiliation with the company until 2011. In 2008, Brooks founded Rethink Robotics based on his belief that existing manufacturing robots were too complicated for regular use in manufacturing. He continues to serve as chief technical officer and chairman of the company today.

Brooks has also long been active on a variety of professional committees and activities. He has served on the International Scientific Advisory Group of the National Information and Communication Technology Australia, the Global Innovation and Technology Advisory Council of John Deere & Company, the Visiting Committee on Advanced Technology of the National Institute of Standards and Technology, and the General Electric Robotics Advisory Council. Since January 2016 he has also served as deputy chairman of the Advisory Board of Toyota Research Institute, an organization actively involved in robotics research.

Brooks has also been elected to a number of scientific institutions and organizations, such as the National Academy of Engineering, the Association for the Advancement of Artificial Intelligence (of which he was a founding member), the American Academy of Arts & Sciences, the American Association for the Advancement of Science, the Association for Computing Machinery, the Institute of Electrical and Electronics Engineers, the Australian Academy of Science, and the Australian Academy of Technological Sciences and Engineering.

Brooks has been awarded the Computers and Thought Award of the International Joint Conference on Artificial Intelligence in 1991, the IEEE Inaba Technical Award for Innovation Leading to Production in 2008, the Robotics Industry Association's Engelberger Robotics Award for Leadership in 2014, and the IEEE Robotics and Automation Award in 2015. He has received honorary doctorates from Flinders University in 2016 and Worcester Polytechnic Institute in 2017.

George Devol (1912–2011)

Devol is regarded by many experts in the field as the father of the industrial robot. In the early 1950s, he had become interested in automating the process by which objects in factories can be lifted and moved more efficiently by mechanical devices than by humans. He had explained to a friend on one occasion that, after all, "50 per cent of the people in factories are really putting and taking." He thought why valuable human labor should be devoted to such mindless activities when machines could probably do the job more efficiently. Devol was living in a time when the technology for just such devices, electrical motors, generators, and transmission systems; programs for systematic control of operations; and radio (and soon television) technology were just being developed. His vision of a mechanical device to automate routine operations was entirely within the scope of such developments.

George Charles Devol Jr. was born in Louisville, Kentucky, on February 20, 1912. He attended Riordan Preparatory School but decided not to continue with high school or college. He was, in fact and to a large extent, self-educated in the fields of electricity and mechanics in which he was to make his mark as an inventor. Instead of pursuing a typical education, he went to work for a series of electronics companies until, in 1932, he founded his own business, United Cinephone. The company was created to find ways of dealing with a new problem in the entertainment industry, the coordination of sound and picture in motion pictures. The first "talkie" films had just begun to appear at the time, and one of the first problems producers encountered was keeping the spoken word in the film in phase with actors' facial movements. Devol saw an opportunity to make a breakthrough with his inventions, but was ultimately unsuccessful. The company went out of business in 1943. During its period of operation, United Cinephone diversified, developing a number of new products, such as a lighting system for garment factories called Orthoplane, an optical registry system for color printing presses, and phonographs and amplifiers. In 1939, Devol developed a photoelectric device for counting individuals as they passed through the entry gate at the New York World's Fair.

Undeterred by his lack of success with "talkies," Devol began to apply his knowledge and skills to another problem, eventually creating the "Phantom Doorman," a door that detected the presence of a person with a photocell and responded by opening a door for the purpose. The invention is essentially the same system that is used for automatic doors today.

The late 1940s was also a period of growing interest in robotics. To a considerable extent, that interest grew out of a short story by Isaac Asimov, "Liars," in which the author first coined the term *robotics*. Devol was already imagining the use of mechanical devices (especially mechanical arms), capable of picking up and moving heavy objects during manufacturing operations. In 1954, he applied for a patent for a device he called Programmed Article Transfer. The device was, to a large extent, the basis for most modern industrial robots.

While attending a cocktail party in Westport, Connecticut, in 1956, Devol met an individual whose interests and skills complemented his own, Joseph Engelberger. Unlike Devol, Engelberger had earned a degree in engineering at Columbia University and was currently working as chief of engineering at the firm of Manning, Maxwell, and Moore in Stratford, Connecticut. The two men were brought together at an informal get-together because of their common interest in robotics. Together they obtained financing from the firm of Consolidated Diesel Electronic and formed a firm for the production of a robotic arm, a company called Unimation Incorporated.

The company's first sale of a robotic device, called a Unimate, was made to General Motors, which installed the machine on an assembly line at its plant in Ewing Township, New Jersey. The robotic arm was used to lift and stack die-cast parts and to do spot welding. Before long, other car manufacturers, such as Ford and Chrysler, began to understand the advantages of mechanism of their production lines, and Unimation company began supplying Unimate machines with other capabilities, such as spray painting, welding, and application of adhesives.

This line of "progress" was greeted enthusiastically by production companies, who appreciated the speed and efficiency with which the machines worked. They were less well accepted by workers on the line who saw their jobs being taken over by machines, with the prospect of that process accelerating in the future, a controversy that remains in the field of robotics today.

Devol was inducted into the Inventors Hall of Fame in 1997. At the induction ceremony, a robot called Sico walked up to him and said "Father, so good to see you!" In spite of his success with the Unimate, Devol never tired of looking for new fields of invention. For example, he developed a precursor of the modern microwave oven with the production of a device called the Speedy Weenie, a device that cooked and delivered a hot dog by way of a vending machine. Such facilities were made available in public areas, such as the Grand Central railway terminal in New York City (photo of the machine available at https://www.jameco.com/jameco/pressroom/devol.html). Devol died at the age of 99 on August 11, 2011, in Wilton, Connecticut.

Joseph Engelberger (1925–2015)

Joseph Engelberger has been called the father of robotics, a title often given to his business partner and inventor, George Devol. Although Devol developed the technology on which the first industrial robot was based, Engelberger provided the managerial and financial support needed to turn Devol's invention into a practice commercial device. Engelberger and Devol met at a cocktail party in Westport, Connecticut, in 1956, shortly after Devol had applied for a patent for his new device. Although he originally called the machine Programmed Article Transfer, Engelberger said that it "sounded like a robot" to him. The conversation went well because of Engelberger's and Devol's shared background in engineering and their common interest in science fiction. Devol's invention sounded to both of them as if one of the standard creatures of science fiction had finally been created. And it was a device that was more than a subject of novels, short stories, and motion pictures; it was a device that could actually be put to use in industrial operations, such as picking up and moving packages, the action from which it got its original name.

Engelberger's role in the partnership was to obtain financing for the commercial development of the new machine. For that task, he turned to the company for which he was working at the time, Condecs, an industrial engineering firm. With that funding, production of the robot began in earnest, and the first model was sold to General Motors for use in its die-casting plant in Trenton, New Jersey. By that time, the machine's name had been changed from Programmed Article Transfer to Unimate, named after the Devol–Engelberger company's name Unimation, for universal automation. The first Unimate cost $65,000 to produce and was sold for $18,000. The Engelberger–Devol gamble turned out to be a smart move, however, as General Motors soon ordered an additional 66 models, and overseas interest in the robot developed rapidly. In recognition of its special place in the history of robotics, Unimate was inducted into the Robot Hall of Fame in 2003.

Joseph Frederick Engelberger was born on July 26, 1925, in Brooklyn, New York. While still a child, his family moved to Connecticut, where they were to live throughout the difficult days of the Great Depression. During this period, Engelberger immersed himself in science fiction, with the great author, Isaac Asimov, being his favorite writer. His reading experience convinced him that he was interested in more than just *reading* about robots; he also wanted to build them.

Engelberger's immediate future was set for him in 1942, at the beginning of World War II. Given the somewhat desperate situation in which the military found itself, it lowered the draft age to 18, making Engleberger subject to the draft. When called, he was assigned to a special program called the V-12 Navy College Training Program at Columbia University, where he studied some of the latest developments in engineering technology. At the completion of his training, he served in the navy for a brief period of time; he was transferred to the Bikini Atoll in the Marshall Islands, where he was assigned to work on the atomic bomb. At the war's conclusion, he returned to Columbia, where he was awarded his bachelor's degree in physics in 1946 and his master's degree in electric engineering in 1949.

After completing his MS at Columbia, Engelberger joined the manufacturing firm of Manning, Maxwell, and Moore, a company with a long corporate history in the United States. He put to work his knowledge of physics and nuclear engineering at the company, where his assignment covered the development of controls for jet airplanes and nuclear power plants. When the company later merged with Dresser Industries in 1964, Engelberger found himself out of a job, a happenstance that came at just the right time for his decision to work with Devol on the Unimate.

After succeeding with industrial robots, Engelberger began to imagine broader applications of the technology, specifically as helpers in the home (domestic robots) and in the field of health care. In 1984, he founded the Transitions Research Corporation for the purpose of building and selling medical robots. The first product in this line was called HelpMate. It was able to transit hospital hallways and patient rooms, much as similar robots are able to do today, automating and increasing the efficiency of a number of hospital procedures. He continued to work in the field of robotics, always looking for new applications for the machines, until his death in Newtown, Connecticut, on December 1, 2015, about four months after his 90th birthday.

Engelberger received the 1982 Nyselius Award from the American Die Casting Institution, 1982 Leonardo da Vinci Award of the American Society of Mechanical Engineers, 1982 American Machinist Award, the Progress Award of the Society of Manufacturing Engineers in 1983, Golden Omega Award at the Electrical Electronics Insulation Conference, 1983

McKechnie Award from the University of Liverpool, 1984 Egleston Medal from Columbia University, 1997 Beckman Award for pioneering and original research in the field of automation, 1997 Japan Prize, and IEEE Robotics and Automation Award in 2004. He was elected to the National Academy of Engineering in 1984. In 1977, the Robotics Industries Association created the Joseph F. Engelberger Awards, given to "persons who have contributed outstandingly to the furtherance of the science and practice of robotics."

Future of Humanity Institute

The Future of Humanity Institute (FHI) was founded in 2005 by Nick Boström, who was also instrumental in the founding of the Institute for Ethics and Emerging Technologies. FHI was originally funded by the James Martin 21st Century School, whose name was later changed to the Oxford Martin School. The school claims to be the only one of its type in the world, with more than 200 researchers working on "the most pressing global challenges and opportunities of the 21st century." The research in which FHI is involved has been described by Boström in a now somewhat famous article in the November 23, 2015, issue of *The New Yorker*, entitled "The Doomsday Invention," with the subhead "Will artificial Intelligence Bring Us Utopia or Destruction?" (The article is available at https://www.newyorker.com/magazine/2015/11/23/dooms day-invention-artificial-intelligence-nick-bostrom. Accessed on October 30, 2017.)

As of late 2017, the institute consists of about a dozen members of the research staff, about a half dozen research associates, and support and administration teams. The research team divides its efforts into four major categories: macrostrategy, AI safety, governance of AI program, and biotechnology. The term *macrostrategy* refers to the ways in which the decisions that humans make today are likely to influence the type of world we will live in in the future. Research in this area has produced papers dealing with topics such as comparisons of predictable and unpredictable lives; the paths, dangers, and strategies associated with superintelligence; global catastrophic risks; anthropic bias; and the likelihood of a doomsday catastrophe.

The potential risks that artificial intelligence (AI) may pose to human civilization are a second area of research. Some specific issues that are being addressed by FHI staff are how AI systems can learn safely in the real world, a study of the values that humans hold and how this relates to "thinking machines," and what range and limit AI machines might have. Governance of AI program is concerned with the policies and actions that nations, industry, and individuals might formulate to increase the likelihood that AI will remain essentially under human control. Some recent papers in this area deal with the question of whether and/or when AI systems will ever become more "intelligent" than humans and what such an event might mean for society; what some of the requirements might be to retain control of machines with superintelligence; and what some unprecedented technological risks may arise in the future. Specific topics in biotechnology mirror many of the research interests in the scientific community at large, involving a consideration of the possible biotechnological agents that might be developed and the threats they might pose to human societies.

A complete list of all publications produced by the institute, along with links to the actual documents, is available at https://www.fhi.ox.ac.uk/publications/.

Future of Life Institute

The Future of Life Institute (FLI) was founded in 2014 by a group of concerned scientists and other interested individuals "to catalyze and support research and initiatives for safeguarding life and developing optimistic visions of the future, including positive ways for humanity to steer its own course considering new technologies and challenges." The founders included Jaan Tallinn, cofounder of Skype; Max Tegmark, professor of physics at the Massachusetts Institute of Technology; Viktoriya Krakovna, research scientist in AI safety at DeepMind; Anthony Aguirre, professor of physics at the University of California, Santa Cruz; and Meia Chita-Tegmark, PhD candidate at Boston University. The organization has also formed an advisory board that consists of a number of esteemed individuals from a variety of other fields, such as actors Alan Alda and Morgan Freeman and researchers such as Stephen Hawking, Nick Boström, Elon Musk, and Martin Rees.

Research at FLI is currently focused in four general areas: artificial intelligence (AI), biotechnology, climate change, and nuclear power. The institute issues occasional reports and new stories about the status of progress in each of the areas, along with the potential threat that such advances pose. One basic publication, as an example, is an online essay on benefits and risks of artificial intelligence. It discusses safety issues associated with AI, "top myths" about AI, timeline predictions, potential risks of superhuman AI, and some fundamental controversies about the field. This publication is of special importance because of the extensive list of references, with their links, that deal with these issues.

The section on biotechnology carries similar reports and commentaries on progress in biotechnology. Some recent items listed current risks of biotechnology inventions, GP-write (a genome project), a review of events from 2016, artificial photosynthesis, federal biotechnology regulations, and the way in which earthquakes could destroy humanity. Current developments in the field of nuclear power, especially nuclear weapons, and climate change are also reviewed on the FLI Web site.

Thus far, most events sponsored by FLI are conventions and other meetings, as well as lectures by members of the organization and other interested individuals. For example, it made possible a question-and-answer podcast by Max Tegmark on his new book *Our Mathematical Universe*; a speech by Nick Boström on "Superintelligence—Paths, Dangers, Strategies"; a symposium on "Neural Information Processing Systems"; and a talk by Martin Rees on "Catastrophic Risks: The Downsides of Advancing Technology." The association held its first annual general conference on "The Future of AI: Opportunities and Challenges" on January 2–5, 2015, in San Juan, Puerto Rico.

Heron of Alexandria (ca. 10–70 CE)

The birthdate of Heron (also Hero) of Alexandria is not known with certainty. The best guess available is based on a report he wrote in 63 CE of an eclipse of the moon. Little is known also of his life or career, although evidence suggests that he worked and taught at the Musaeum (also Mouseion) in Alexandria. The Musaeum was affiliated with one of the most famous institutions of the ancient world, the Library at Alexandria, established in the third century BCE as a repository for the greatest writings of the ancient world. Evidence suggests that the body of Heron's writings was originally prepared as lecture notes. Little is known about his personal life or about his research. Much of that which remains comes down to the present day in later writings by Heron's followers, on whom he had a strong influence.

Heron is probably best known today for his inventions, many of which included some of the earliest automata of which we have a record. One such device was an organ that played on its own, powered by the flow of air powered by a windmill. The organ is sometimes said to be the first wind-powered machine to have been invented. Another famous device was an automaton regarded as the world's first vending machine. Insertion of a coin into the top of the machine prompted the release of a small amount of holy water. Heron apparently also produced automata for the Greek theater, perhaps the most famous of which was a puppet play that ran for about 10 minutes, powered by a complex system of ropes, pulleys, and other simple machines. As background, Heron devised a method for producing thunder-like sounds, obtained by the mechanical release of a metal ball onto a drum.

In addition to his automata, Heron invented a variety of other machines and discovered some important physical principles. For example, he is credited with the creation of the first force pump, a device consisting of two opposing pistons operating in concert to produce a constant flow of water. The principle was later adapted by the Romans for use as the world's first fire engine. Heron is also credited with the discovery of some basic principles of optics that were, in fact, largely ignored for centuries after his death.

Much of that which is known about Heron's work is preserved in his two volumes, *Pneumatica* and *Mechanica*. The former book summarizes much of what was then known and all of which Heron had discovered about the movement of gases. The operation of many of his automata was described in the former book, while the latter was intended as a working guide for the movement of heavy objects for architects and engineers. *Mechanica* also includes the descriptions of some of Heron's automata. Other works attributed to Heron include "Dioptra" (a book on land surveying), "Metrica" (methods for calculating surface areas and volumes), "Definitiones" (definitions of terms from geometry), "Geodaesia" (construction of siege machines), "Belopoeica" (construction of war machines), "Catoptrica" (propagation of light, its transmission, and reflection), and "Geometrica" (the principles of geometry).

IEEE Robotics & Automation Society

The IEEE Robotics & Automation Society (IEEE RAS) was founded in December 1983 as the IEEE Robotics & Automation Council (IEEE RAC). Within the administrative structure of the IEEE (Institute of Electrical and Electronics Engineers), the terms *council* and *society* have quite different meanings. An IEEE council is made up of organizations with interests similar to those of the parent organization, while an IEEE society consists of individual IEEE members. The founding associations from which IEEE RAC was created were Aerospace and Electronic Systems; Circuits and Systems; Components, Hybrids, and Manufacturing Technology; Computers; Control Systems; Industrial Electronics; Industry Applications; and Systems, Man, and Cybernetics. The motivation for the creation of the original organization was a significant increase in the number of papers, tutorials, notices of meetings, and other formats being received by professional journals in the field. A group of IEEE members decided that this growing interest justified the creation of a distinct council focused specifically on robotics issues. Six years later, the council decided to change its format and open its membership to all members of IEEE, thus becoming a society rather than a council. The appropriate name change occurred in 1987.

IEEE RAS currently has about 13,000 members in 120 countries around the world. There are 4,392 regular members in the United States, of whom 350 are student members. India has the second-largest number of members (894) and Japan the third largest (778).

IEEE RAS is managed by an administrative committee that consists of the association president, 18 members at large, and 6 members of the executive committee, each of whom is responsible for a distinct board dealing with some aspect of the organization's work. Those boards are concerned with conference activities, financial activities, industrial activities, member activities, publication activities, and technical activities. Another set of standing committees are responsible for other aspects of IEEE RAS work: Advisory Committee, Awards Committee, Constitution and Bylaws Committee, Long-Range Planning Committee, Nominations Committee, Robotics and Automation Research and Practice Ethics Committee, and Special Interest Group on Humanitarian Technologies.

On another level, IEEE RAS maintains more than 40 technical committees organized to deal with specific problems in the field of robotics. A list of such committees currently includes Aerial Robotics and Unmanned Aerial Vehicles, Agricultural Robotics and Automation, Automation in Health Care

Management, Autonomous Ground Vehicles and Intelligent Transportation Systems, Bio Robotics, Cognitive Robotics, Cyborg and Bionic Systems, Human–Robot Interaction and Coordination, Marine Robotics, Multi-Robot Systems, Rehabilitation and Assistive Robotics, Robot Ethics, Robot Learning, Robotics and Automation in Nuclear Facilities, Space Robotics, Surgical Robotics, Sustainable Production Automation, and Wearable Robotics.

One of the most important parts of IEEE RAS's work is sponsorship and cosponsorship of conferences on robotics. Three major annual conferences are organized around the topics robotics and automation, automation science and engineering, and intelligent robots and systems. In all, the organization sponsors on its own 10 annual conferences, including the three mentioned earlier, financially cosponsors 17 conferences a year, and sponsors and cosponsors 30 other technical conferences and workshops. Examples of the topics of some of these meetings are advanced robotics and its social impacts; humanoid robots; microelectromechanical systems; safety, security, and rescue robotics; haptics (nonverbal communication); and assembly and manufacturing.

IEEE RAS conducts an especially active program of publications on robotic topics. It sponsors 4 major publications (*IEEE Robotics and Automation Letters, IEEE Robotics and Automation Magazine, IEEE Transactions on Automation Science and Engineering,* and *IEEE Transactions on Robotics*) and cosponsors 15 other publications, such as *IEEE/ASME Transaction on Mechatronics, IEEE Journal of MicroElectroMechanical Systems, IEEE Transactions on Haptics, IEEE Transactions on Intelligent Vehicles, IEEE Transactions on Nano-BioScience, IEEE Transactions on Cognitive and Developmental Systems, IEEE Sensors Journal,* IEEE Systems Journal, *IEEE Transactions on Affective Computing, IEEE Transactions on Consumer Electronics,* and *IEEE Transactions on Control of Network Systems.*

Educational programs also constitute an essential part of the IEEE RAS agenda. For example, the organization produces webinar and video programs dealing with topics of special interest to members, such as surgical robots and haptics. It also sponsors and cosponsors a number of robotic competitions among students of all ages, designed to encourage interest in the field and promote STEM (science, technology, engineering, and mathematics) programs. IEEE RAS also sponsors a Distinguished Lecturer program that provides specialists in the field who visit local associations and other groups to talk about their field of interest. It also cosponsors, with Indiana University, a Robotics History project. It has also developed a variety of education materials, available online, for use in schools and other educational settings. Some topics of projects currently available are "Case Study of Modeling and Control in Energy-Efficient Buildings," "The Robot Programming Network," "Robot Programming for All," "Development of an Underwater Robotics Research Training Program," "Professional Education in Service Robotics," "Try-a-Bot," "Drones Demystified!" and "RobotCup@Home Education Open Courseware for Service Robotics."

Institute for Ethics and Emerging Technologies

The Institute for Ethics and Emerging Technologies (IEET) was founded in 2004 by philosopher Nick Boström and bioethicist James J. Hughes. The organization is a nonprofit "think tank" focusing on a topic its calls "technoprogressive technologies." The term was created to describe the promotion of technologies that can have positive effects on human society, such as freedom, happiness, and self-fulfillment. The major questions in which IEET is interested are as follows:

- Which technologies, especially new ones, are likely to have the greatest impact on human beings and human societies in the 21st century?
- What ethical issues do those technologies and their applications raise for humans, our civilization, and our world?

- How much can we extrapolate from the past and how much accelerating change should we anticipate?
- What sort of policy positions can be recommended to promote the best possible outcomes for individuals and societies? (https://ieet.org/index.php/IEET2/about)

One mechanism by which IEET carries out its mission is a series of local lectures and meetings on specific topics relating to its goals. Some of the topics discussed in 2017 include "Beyond Humanism"; "Longevity and Cryopreservation"; "Humans, Machines, and the Future of Work"; "Technological Singularity"; "Cyborg Law"; and "Healthy Aging." The association publishes a (mostly) weekly newsletter "Ethics and Emerging Technologies," available to interested individuals at http://ieet.us12.list-manage2.com/subscribe?u=6c2c3323a6978d3868f6c4f9c&id=684579a215. It also produces a bimonthly scholar journal, *Journal of Evolution and Technology*. Fellows and staff of IEET have also published a number of books on technoprogressive technologies, such as *Conversations with the Future: 21 Visions for the 21st Century, Surviving the Machine Age: Intelligent Technology and the Transformation of Human Work, eHuman Deception, Free Money for All: A Basic Income Guarantee Solution for the Twenty-First Century, Humans and Automata: A Social Study of Robotics*, and *Ten Years to the Singularity If We Really Really Try*. Finally, IEET sponsors a blog that focuses on ethical issues of emerging technologies, https://ieet.org/index.php/IEET2/IEETblog.

International Federation of Robotics

The International Federation of Robotics (IFR) was founded in 1987 as a nonprofit organization that presently consists of more than 50 companies from more than 20 nations. The purpose of the organization is to promote research, development, use, and international cooperation in the entire field of robotics, including primarily industrial and service robots. Some members of IFR are ABB Robotics (Sweden), Blue Ocean Robotics (Denmark), CLOOS (Germany), Daihen (Japan), and 3M (United States). Some of the trade organizations that are also members include Asociación Española de Robótica (Spain), British Automation & Robotics Association (United Kingdom), China Robot Industry Alliance, Danish Industrial Robot Association, Japan Robot Association, Associazione Italiana di Robotica e Automazione (Italy), Russian Association of Robotics, and Robotic Industries Association (United States).

One of IFR's primary activities is sponsorship of the annual Innovation and Entrepreneurship in Robotics and Automation. The award is given to individuals who have contributed notable basic research in the field of robotics or promoted the incorporation of such research into concrete applications. The awards are one of the ways in which IFR carries out its objective of helping to make the industrial world aware of the latest developments in robot research and to incorporate that research into its own activities.

Another major activity is the collection of statistics for the production, distribution, and research of robots in the world. These data are provided in the organization's annual "World Robotics" report, sections of which can be downloaded at https://ifr.org/free-downloads/. IFR also sponsors the International Symposium on Robotics, an annual meeting that was first held in 1970 and has been repeated every year since. The symposium provides an opportunity for researchers and entrepreneurs in the field to get together and share information about recent developments in the field.

The association is a primary source of information on the state of robotics worldwide. It makes that information available to the industry, to the press, and to the general public, not only through the symposium but also through a number of online articles on the topic and through position papers on the topic of robotics. An interesting recent example of the latter is a paper entitled "The Impact of Robots on Productivity, Employment and Jobs," which summarizes recent research on the effects of the increasing use of robots on employment and the workplace. The paper is available at https://ifr.org/img/office/IFR_The_Impact_of_Robots_on_Employment.pdf.

Hiroshi Ishiguro (1963–)

Hiroshi Ishiguro has created a series of robots with strongly human-like features. He may perhaps best be known for one of his androids that is a virtually identical copy of himself. He has also made a copy of his four-year-old daughter and a number of other unspecified androids and gynoids. His purpose in pursuing this line of research is not only to develop human-looking robots, which may someday have a host of tasks to perform in human homes, but also to learn more about the way the human brain works. This information will be of particular importance in the rapidly growing field of human–robot interactions. He is currently director of the Intelligent Robotics Laboratory of Osaka University, in Osaka, Japan.

Hiroshi Ishiguro was born in Shiga Prefecture, just east of Kyoto, Japan, on October 23, 1963. Although he was very much interested in art as an undergraduate at Yamanashi University, he eventually committed to a major in robotics and received his engineering degree in 1986. He then continued his studies at Yamanashi, where he received his master's degree in 1988, and then at Osaka University, from which he earned his doctor of engineering degree in the field of systems engineering in 1991.

Ishiguro then stayed on at Yamanashi University as a research associate from 1992 through 1994 before accepting an appointment as associate professor in the Department of Social Informatics at Kyoto University. He then became associate professor (2000–2001) and professor (2001–2002) in the Department of Computer and Communication Sciences at Wakayama University. In 2002 he was named professor in the Department of Adaptive Machine Systems in the Graduate School of Engineering Science at Osaka University (2002–2009). He was then promoted to his current position of professor in the Department of Systems Innovation in the Graduate School of Engineering Science at Osaka University. Other appointments he has held include visiting researcher (1999–2002) and visiting group leader (2002–present) of the Intelligent Robotics and Communication Laboratories at the Advanced Telecommunications Research Institute and visiting scholar at the University of California at San Diego (1998–1999).

Ishiguro has written more than 600 scholarly papers in subjects such as what is a human?, an android counselor, a robot with social intelligence, does a conversational robot need to have its own system of moral?, intimacy of robotic conversations on the telephone, the effects of a robot's touch on human behavior, robotic interventions for autistic children, infants' perceptions of robotic presence, design of an android with polite behavior, and nodding responses in human-like robots.

Among the awards Ishiguro has received are the Best Humanoid Award (kid size) at RoboCup 2006 in Bremen, Germany; Best Paper and Poster Awards at the Second ACM/IEEE International Conference on Human-Robot Interaction (2007); Best Paper Award at the Fourth ACM/IEEE International Conference on Human-Robot Interaction (2009); and Prize for Science and Technology by the Minister of Education, Culture, Sports, Science, and Technology, Japan (2015).

Al-Jazari (1136–1206)

Al-Jazari was one of the best-known and most highly esteemed scholars of his era. He is perhaps best known for his book on mechanics, *Al-Jami 'bayn al-'ilm wa 'l-'amal al-nafi 'fi sina 'at al-hiyal (A Compendium on the Theory and Useful Practice of the Mechanical Arts)*, which dates to 1206, the year of al-Jazari's death. It has been described by one reviewer as "the most comprehensive and methodical compilation of the current knowledge about automated devices and machines." The book contains detailed descriptions not only of such devices known and used throughout the known world at the time but also of the many mechanical marvels that al-Jazari himself invented.

Al-Shaykh Ra'is al-A'mal Badi'al-Zaman Abu al-'Izz ibn Isma'il ibn al-Razzaz al-Jazari, his full Islamic name, is thought to have been born in 1136 in Cizre, a city in the state of Artuqid, in the eastern region of modern-day Turkey, adjacent to the border of modern-day Iraq. Nothing is known about his early life. As an adult, al-Jazari served in the same office as his father, chief engineer at the Artuqid court. According to his own autobiographical sketch in *Al-Jami*, al-Jazari was asked by the Artuqid imam (ruler), Nur al-Din Muhammad, to write the book. At that point, al-Jazari says that he had joined the court in 1174 and had already spent 25 years in service to the imam at the time he requested the book to be written. Al-Jazari spent the rest his life working on the book, completing it only in the year of his death, 1206.

Al-Jami contains descriptions of about 100 mechanical devices, including water and candle clocks, vessels designed for drinking sessions, pitchers and basins for blood-letting and washing before prayers, fountains with changeable shapes, automata that play the flute, machines for raising water, and miscellaneous devices. Some of the simpler devices described by al-Jazari include a primitive camshaft and crankshaft, a segmental gear, and an escapement mechanism for a rotating wheel.

Among his automata include a waitress who serves drinks to the observer, a hand-washing device with a mechanism for flushing water, a peacock fountain attended by servants, and a musical robot band. The last of these devices consisted of a boat carrying four musicians that played their instruments as they floated across the water. Possibly the most famous of all al-Jazari's inventions was a large clock made in the shape of an elephant. A modern-day replication of that device stands in the Ibn Battuta Mall in Dubai, United Arab Emirates. (For a video of the device, see https://www.youtube.com/watch?v=5QaDuyPz98Q.) In addition to his significant skills as an inventor and engineer, al-Jazari was an accomplished artist. A number of the illustrations included in *al-Jami* were produced by the artist himself.

Al-Jazari died in 1206 at an unknown location.

Ray Kurzweil (1948–)

Kurzweil is an American computer scientist, inventor, futurist, and author. He is one of the most outspoken individuals who are warning of the technological singularity, which he predicts will arrive within a matter of decades. His thoughts on the subject are perhaps best expressed in his 2005 book, *The Singularity Is Near: When Humans Transcend Biology*. He is also a highly respected inventor, who is responsible for the production of a computer-based system for musical composition (1964); a computer-based system for college selection (1967); the first program for converting text to speech (1975); the first such device designed specifically for the blind (1976); the first knowledge-based system for creating medical reports (1985); the first speech recognition dictation system for Windows (1994); the first robot, Ramona, to perform in front of a live audience with a live band (2001); and the first realistic, human-looking chatbot (Ramona) with facial expressions capable of holding a conversation with a human (2001).

Kurzweil has also been a prolific entrepreneur, with the founding of a number of new companies to his credit, including (but not limited to) Kurzweil Computer Products, Inc. (1974); Kurzweil Music Systems, Inc. (1982); Kurzweil Applied Intelligence, Inc. (1982); Kurzweil Technologies, Inc. (1995); Kurzweil Educational Systems, Inc. (1996); FAT KAT (Financial Accelerating Transactions from Kurzweil Adaptive Technologies, 1999); Kurzweil Cyber Art Technologies, Inc. (2000); KurzweilAI.net (2001); Ray & Terry's Longevity Products, Inc. (2003); K-NFB Reading Technology, Inc. (2005); and Kurzweil Capital Partners, LLC (2005).

Raymond ("Ray") Kurzweil was born on February 12, 1948, in Queens, New York. Kurzweil was a precocious child who had made up his mind by the age of five that he wanted to become an inventor. He began his pursuit of this objective with a number of projects carried out at home. While still an elementary school student, for example, he built a robotic puppet theater and robotic game. By the age of 12, he had also become fascinated with computers and was building his own machines at a time when the world at large was only vaguely aware of the role that computers would have in the future. Kurzweil credits his interest in predictions of the future to his home life. His parents were also interested in technology and its potential effects on the future of civilization, so that such discussions were a regular part of the household conversations.

Kurzweil attended Martin Van Buren High School in Queens, where he was more interested in his own inventions and research than the regular class schedules. One of his most notable inventions was completed at the age of 17, when he designed a software program that was able to recognize musical patterns and then convert those patterns into original musical compositions. He won first prize in the International Science Fair in 1967 for his invention.

After graduating from high school, Kurzweil matriculated at the Massachusetts Institute of Technology (MIT), where he took every course in computer programming (nine in all) then available. He was awarded his BS degree in computer science and literature by MIT in 1970 but chose not to pursue postgraduate studies. Indeed, by the time he graduated, he had already created his first commercial product, a computer program designed to help high school students select a college to attend. He sold that company to the publishing firm of Harcourt, Brace & World for $100,000 and royalties. With his degree in hand, Kurzweil was ready to begin his career of invention and writing.

Kurzweil's work has been recognized by a number of important honors and awards. He has been granted honorary doctorates by Babson College, Clarkson University, McGill University, Pennsylvania College of Optometry, Bloomfield College, DePaul University, Worcester Polytechnic Institute, Landmark College, Michigan State University, Dominican College, Queens College of the City University of New York, New Jersey Institute of Technology, Misericordia College, Merrimack College, Rensselaer Polytechnic Institute, Northeastern University, Berklee College of Music, and Hofstra University. Among his other honors are the Leonardo da Vinci Medallion of Honor of the Leonardo da Vinci Society, Arthur C. Clarke Foundation Lifetime Achievement Award, Forum Nokia Americas Tech Showcase Regional Award, American Creativity Association Lifetime Achievement Award, Chicago Lighthouse Centennial Award, National Federation of the Blind Newell Perry Award, IEEE Alfred N. Goldsmith Award for Distinguished Contributions to Engineering Communication, Lifeboat Foundation Guardian Award, Second Annual American Composers Orchestra Award for the Advancement of New Music in America, Lemelson-MIT Prize, National Medal of Technology, President's Award of AHEAD (Association on Higher Education and Disability), Software Industry Achievement Award, Massachusetts Software Council, Access Prize, American Foundation for the Blind, Dickson Prize from Carnegie Mellon University, Gordon Winston Award of the Canadian National Institute for the Blind, ACM Fellow Award of the Association for Computing Machinery, Massachusetts Quincentennial Award for Innovation and Discovery, Louis Braille Award of the Associated Services for the Blind, Engineer of the Year Award of *Design News* magazine, MIT Founders Award, 1986 White House Award for Entrepreneurial Excellence, Francis Joseph Campbell Award of the American Library Association, National Award for Personal Computing to Aid the Handicapped of Johns Hopkins University, and Grace Murray Hopper Award of the Association for Computing Machinery. Kurzweil has also been elected to the AES Technology Hall of Fame, the Electronic Design Hall of Fame, CRN

Industry Hall of Fame, National Inventors Hall of Fame, and the Computer Industry Hall of Fame.

Leonardo da Vinci (1452–1519)

Leonardo da Vinci (commonly known simply as Leonardo) is arguably one of the most famous scholars in the history of mankind. He is perhaps best known as the creator of two famous paintings, the *Mona Lisa* and the *Last Supper*. He was also a student of human anatomy and physiology, the topics of many of his sketches and paintings. He was one of the most imaginative inventors of his, or any, time, producing drawings (and sometimes working models) of machines such as early versions of the helicopter, parachute, and tank. He also made contributions in fields as diverse as architecture, astronomy, botany, cartography, geology, literature, mathematics, music, and sculpture.

Da Vinci's broad range of interests also included the field of automata, where he is most famous for three inventions: a mechanical knight, lion, and self-propelled cart. The mechanical knight was apparently designed to be a full-size model of a warrior clad in a traditional suit of armor. It was capable of standing, sitting, raising its visor, moving its arms, opening and closing its mouth, and moving its head from side to side. Some experts believe that the device was also able to make sounds (although not speech) by an automated drum stored within its body. The knight has fascinated scholars throughout history. Most recently, robotic engineer Mark Rosheim used da Vinci's sketch books to construct a model of the knight, which he displayed and discussed on a BBC program in 2002. (For more about the model, see https://www.wittystore.com/leonardo-da-vinci-robot.)

Da Vinci's mechanical lion was apparently built as a gift for King Francis I of France, to whom it was presented in 1515. The lion moved about like the real animal and concluded its act by presenting a bouquet of flowers to the king. The self-propelled cart seems to have been made for theatrical performances and consisted of a complex combination of springs, wheels, pulleys, and other simple machines.

The story of Leonardo's life and career has been the subject of many formal biographies, the most recent of which was American writer Walter Isaacson's *Leonardo da Vinci* (Simon & Schuster, 2017). He was born Leonardo di ser Piero da Vinci on April 15, 1452, near the town of Vinci (from which he gets the name for which he is best known) in the Republic of Florence. His mother was a peasant woman named Caterina, who had become pregnant by a wealthy legal notary (a type of lawyer), Messer Piero Fruosino di Antonio da Vinci, from the nearby town of that name. Leonardo spent the first five years of life in his mother's home in the tiny town of Anchiano, a short distance from Vinci. He then moved to his father's estate, where he spent the rest of his childhood.

While at his father's home, Leonardo received informal instruction in mathematics and Latin. By the age of 14, however, his skills as an artist had convinced his father to apprentice him to the workshop of the noted Florentine artist Andrea di Cione, better known simply as Verrocchio. In 1472, at the age of 20, Leonardo had qualified as a master of the Guild of St. Luke, an international organization consisting primarily of painters, sculptures, and other artists. Qualified then to establish his own workshop, he decided however to remain with Verrocchio for another four years. By 1478, da Vinci was ready to strike out on his own, and he shortly received commissions for his first two works, an altarpiece for the Chapel of St. Bernard in Florence's Palazzo Vecchio, and a painting, *The Adoration of the Magi*, for the monastery at San Donato a Scopeto.

Neither commission appears to have been completed since Leonardo had decided to accept a position in Milan as a member of the court of Ludovico Sforza, Duke of Milan. His assignment then covered a wide variety of fields, reflecting his growing interest and skills in subjects such as music, painting, architecture, engineering, and sculpture. His first project was to be a 16-foot-tall equestrian statue in honor of the founder of the Sforza dynasty, Francesco. Da Vinci worked on the project for more than a decade, but, for a number of reasons, it was never completed.

Leonardo's brief stay in Milan occurred during an epidemic of plague that devastated the city's population. Having already been through a similar experience in Florence in 1479, Leonardo had thought deeply about ways of dealing with this type of disaster. He had worked out a plan for an "ideal city" that included broad roads for unimpeded travel, canals for transportation purposes and the efficient removal of sewage, and separation of the urban area into "upper" regions, where the elite could be protected from disease, and "lower" regions, where lower-class men and women were left more or less "on their own."

Over the next decade, Leonardo moved on a number of occasions to take up positions in a variety of settings, first to Venice, where he served as a naval architect designing bulwarks to protect the city from invasion; then back to Florence to live with a group of Servite monks; then on to Cesena, where he became a member of the court of Cesare Borgia, son of Pope Alexander VI; then back again to Florence, where he rejoined the Guild of St. Luke; and finally a return to Milan, where he remained until 1508.

During his travel throughout Italy, Leonardo continued to produce a number of significant artistic, scientific, and technical accomplishments. During his stay in Venice in 1499, for example, he designed an outfit that is now regarded as the world's first diving suit. Leonardo believed that soldiers could don such suits, move under water, and attack enemy ships from beneath them.

In 1513, Leonardo moved on once more, this time to serve Pope Leo X, during whose service he designed the mechanical lion presented to the French king in 1516. Francis was so impressed with Leonardo's work that he gave him a chateau at Cloux (now Clos Lucé), where he remained until his death on May 2, 1519.

Machine Intelligence Research Institute

The Machine Intelligence Research Institute (MIRI) was founded in 2000 under the name of the Singularity Institute for Artificial Intelligence (SIAI). The organization was created based on the assumption that smarter-than-human artificial intelligence was being created and would probably become an even more powerful force in the future. Given the continued growth of human population in the future, along with the coincidental problems that growth would bring, founders of MIRI felt that efforts should be made to ensure that future artificial intelligence (AI) programs are as "friendly" to the human population as possible. They referred to such forms of AI as "Friendly AI." Such an approach would be in opposition to the development of AI programs with negative impacts on human society, a direction that could very well lead to the extinction of the human species itself.

For the first decade of its existence, MIRI maintained a close relationship with Singularity University, a think tank dealing with problems associated with evolving AI development, located in Silicon Valley. MIRI and SIAI collaborated to produce a series of Singularity Summits, conferences at which researchers and other interested individuals reported on their research and prospects for the future of AI, along with actions that would be appropriate to deal with such changes. Summits were held annually from 2006 through 2011. In 2013, SIAI severed its connections with Singularity University and changed its name to its current appellation, Machine Intelligence Research Institute. With its new name, MIRI changed its mission somewhat by focusing on the role of mathematics and theoretical computer science in its study of future prospects for AI and robotics.

Some of the general principles that direct MIRI research are the following:

- Over time, the mathematical algorithms on which AI is based will almost certainly become more complex. Such changes will result in a more long-term decision making by forms of AI, and small errors in human programming will lead to more significant problems in the AI performance.

- As computer programs for AI become more complex, machines will be expected to learn more and make more decisions on their own without further human input.

- Catastrophic results can be expected if machines are unable to operate within human expectations if they are not provided with some mechanisms for aligning their own goals with those of their makers.

- AI systems can be expected to behave well when the environment within which they work corresponds to that for which they were designed. But changes in that environment, even if they are modest, can result in machine behavior that is not expected and that can be harmful to humans.

Current papers produced by MIRI researchers are based on these general principles. Some recent examples, from 2017, are "Functional Decision Theory: A New Theory of Instrumental Rationality," "Incorrigibility in the CIRL Framework," "A Formal Approach to the Problem of Logical Non-Omniscience," "When Will AI Exceed Human Performance? Evidence from AI Experts," "Cheating Death in Damascus," "Agent Foundations for Aligning Machine Intelligence with Human Interests: A Technical Research Agenda," and "Toward Negotiable Reinforcement Learning: Shifting Priorities in Pareto Optimal Sequential Decision-Making."

Opportunities for researchers to get together to share and discuss their current work are an important feature of the MIRI program. For example, the organization sponsors frequent regional workshops on specific topics within the field of AI. Some recent workshop topics have been "Machine Learning and AI Safety Agent"; "Foundations and AI Safety"; "Logic, Probability, and Reflection"; "Preference Specification"; and "Robustness and Error-Tolerance." MIRI also sponsors an online forum at which researchers can more informally present their ideas about specific topics within the field. Finally, the organization encourages interested individuals to organize their own regional workshops at which they can meet and share research projects at regular intervals. Past such workshops have been conducted in locations such as Berlin; Canberra; Columbus, Ohio; Kagoshima, Japan; Prague; Seattle; Tel Aviv; and Zagreb.

Marvin Minsky (1927–2016)

Minsky is often called the father of artificial intelligence. He was involved with a study of robotics throughout most of his career and was responsible for the invention of a number of robotic systems, such as robotic arms and grippers, computer vision systems, and autonomous learning devices. He was long interested in the development of mechanical devices that were able to mimic the neural functions of the human brain. In fact, one of his earliest inventions (with colleague Dean Edmonds) was a neuroncomputer called Stochastic Neural Analog Reinforcement Computer. The device accepted visual images, which it then translated into movements similar to those exhibited by a rat moving through a maze. It consisted of 40 neuron-like junctions that "learned" how to solve the maze.

Another one of Minsky's invention was the so-called tentacle arm, designed to replicate the structure and movement of a crayfish arm. It was one of a number of robotic devices invented by Minsky, often with colleagues, to perform some function normally carried out by some part of the human body. By the 1970s, Minsky had developed a theory known as the Society of Mind, an attempt to find ways by which some set of mechanical devices can be combined to function in ways similar to human brain function. He brought those ideas together in a book by that name, intended as an introduction to the theory for the average reader. For many years, Minsky also taught a course at the Massachusetts Institute of Technology (MIT), the title of which was also "The Society of Mind." (Course materials are available at https://ocw.mit.edu/courses/electrical-engineering-and-computer-science/6-868j-the-society-of-mind-fall-2011/. Accessed on November 4, 2017.)

Marvin Lee Minsky was born in New York City on August 9, 1927. His father was Henry Minsky, chief of ophthalmology at Mount Sinai Hospital, and his mother was Fannie Reiser, a social activist and Zionist. Minsky attended the Ethical Culture Fieldston School (also known simply as Fieldston) in Manhattan. He then continued his studies at the Bronx High School of Science, arguably the most famous American secondary school devoted primarily to the field of science, and then Phillips Andover Academy, from which he graduated in 1944. He was then drafted into the United States Navy, where he served until the end of the war a year later. At that point, Minsky matriculated at Harvard University, from which he received his BA in mathematics in 1950. He then continued his studies at Princeton University, where he was awarded his PhD in mathematics in 1954.

After graduation from Princeton, Minsky returned to Harvard for three years, where he had the title of junior fellow in the Harvard Society of Fellows. In 1957, he accepted an appointment as a staff member at MIT's Lincoln Laboratory, the first in a series of posts he held at the institute over the next 50 years. In 1958, he became assistant professor of mathematics at MIT and, a year later, founded the MIT Artificial Intelligence Project. He served as codirector of that project from 1959 through 1974, after which he was appointed professor of electric engineering for one year. He then became Donner Professor of Science, a post he held until 1989. He was then appointed to the post of Toshiba Professor of Media Arts and Sciences. After his retirement, he remained on the MIT faculty in an emeritus status. He died in Boston on January 24, 2016.

Among Minsky's many honors and awards were the Turing Award of the Association for Computing Machinery (1970), MIT Killian Award (1989), Japan Prize (1990), Research

Excellence Award of the International Joint Conference on Artificial Intelligence (1991), Joseph Priestly Award (1995), Rank Prize of the Royal Society of Medicine (1995), Computer Pioneer Award of the IEEE Computer Society (1995), R. W. Wood Prize of the Optical Society of America (2001), Benjamin Franklin Medal of the Franklin Institute (2001), and In Praise of Reason Award from the World Skeptics Congress (2002). He was awarded honorary doctorates from the Free University of Brussels in 1986 and Pine Manor College in 1987.

Martin Rees (1942–)

Progress in robotics has long been accompanied by the expression of concerns that such devices would eventually "get out of control," take over the world, and perhaps bring about the end of the human race. That scenario has been the basis of any number of plays, novels, short stories, and other forms of literature. One response to such concerns has been "Oh, that's being overly pessimistic (or harsher). Humans will never cede the Earth to mechanical devices." Therefore, it is of some significance that some of the most highly respected researchers of the modern day have begun to develop reasoned arguments that such scenarios are anything *but* unnecessary warnings. One of the leaders of the campaign for humans to become more watchful about the progress of robotics has been Martin Rees, currently Astronomer Royal for Great Britain, a post he has held since 1995. In one interview, Rees was quoted as saying, "If we look into the future, then it's quite likely that within a few centuries, machines will have taken over—and they will then have billions of years ahead of them. In other words, the period of time occupied by organic intelligence is just a thin sliver between early life and the long era of the machines" (https://theconversation.com/aliens-very-strange-universes-and-brexit-martin-rees-qanda-75277. Accessed on October 28, 2017). Rees was also one of the coauthors of a letter also signed by such luminaries as Stephen Hawking, Elon

Musk, Peter Norvig, Stuart Russell, Martha E. Pollock, and Oren Etzioni. (The letter can be accessed at https://futureoflife .org/open-letter-autonomous-weapons/. Accessed on October 28, 2017.)

Martin John Rees was born in York, England, on June 23, 1942, in the midst of World War II. Both of his parents were forced to move a number of times because of wartime conditions. Eventually the family settled in a rural part of Shropshire, in the English Midlands. There his parents established Bedstone College, a boarding school for children and teenagers between the ages of 4 and 18. The school remains in existence today. Rees himself attended Bedstone to the age of 13, at which point he transferred to the Shrewsbury School for the remainder of his secondary education. He then enrolled at Trinity College, Cambridge, where he majored in mathematics, earning his bachelor's degree in 1963. He then continued his studies at Cambridge, earning both an MS and a PhD in mathematics in 1967.

During his academic career, Rees has served in a number of positions, including research fellow at the California Institute of Technology (Caltech, 1968); fellow at Jesus College, Cambridge (1967–1969); staff member at the Institute of Theoretical Astronomy, Cambridge (1967–1972); senior research fellow at King's College, Cambridge (1969–1972); member of the Institute for Advanced Study, Princeton, New Jersey (1969–1970); visiting associate professor at Caltech (1971); visiting professor at Harvard University (1972, 1988–1989); professor at Sussex University (1972–1973); Gresham professor of astronomy at Gresham College, London (1976–1977); visiting professor at the Institute for Advanced Study, Princeton (1982, 1996, 1997); and Regents visiting fellow of Smithsonian Institution (1984–1988). Throughout this period, Rees also made his academic "home" at Cambridge, where he served as Plumian Professor of Astronomy and Experimental Philosophy (1973–1991), director of the Institute of Astronomy (1977–1982, 1987–1991), professorial fellow at King's College

(1973–1991), Royal Society Professor (1992–2003), official fellow at King's College (1992–2003), professor of cosmology and astrophysics (2002–2009), and Master of Trinity College (2004–2012). Rees also served as president of the Royal Society from 2005 through 2010 and was raised to a peerage as Baron Rees of Ludlow in 2005.

Rees has been made a member or fellow of more than a dozen national scientific societies, including those of Finland, India, Japan, Norway, Russia, Sweden, the Netherlands, Turkey, and the United States. He has also been given more than 30 major honors and awards, including the Gold Medal of the Royal Astronomical Society, Guthrie Medal and Prize of the British Institute of Physics, Balzan Prize, Robinson Prize for Cosmology, Bruce Medal of the Astronomical Society of the Pacific, Bower Prize and Award for Science of the Franklin Institute, Rossi Prize of the American Astronomical Society, the Cosmology Prize of the Gruber Foundation, the Einstein Award of the World Cultural Council, Faraday Award of the Royal Society, Crafoord Prize of the Royal Swedish Academy, Niels Bohr Medal (UNESCO), Templeton Prize, Newton Prize of the Institute of Physics, Paczynski Medal, Dirac Medal and Prize, Nierenberg Prize, and Order of the Rising Sun (Gold and Silver Star). He has written nine books in the field of astronomy and cosmology, among them *Cosmic Coincidences: Dark Matter, Mankind, and Anthropic Cosmology* (with John Gribbin), *Gravity's Fatal Attraction: Black Holes in the Universe*, *Before the Beginning—Our Universe and Others*, *Just Six Numbers: The Deep Forces That Shape the Universe*, and *Our Final Hour: A Scientist's Warning: How Terror, Error, and Environmental Disaster Threaten Humankind's Future in This Century—On Earth and Beyond and What We Still Don't Know*.

Robotics Industries Association

The Robotics Industries Association (RIA) is a trade organization consisting of robot manufacturers, suppliers, consultants, users, and other groups interested in robotics. It claims to be the largest such group in North America. RIA was founded in 1974 in an effort to "drive innovation, growth, and safety in manufacturing and service industries through education, promotion, and advancement of robotics, related automation technologies, and companies delivering integrated solutions." RIA currently has nearly 300 members, of whom about three-quarters are from the United States, 15 percent from Canada, 7 percent from Asia, and 3 percent from Europe. Members are classified into five major groups: robot suppliers, robot system integrators, robot users, robotics consultants and affiliates, and robotics educators and researchers. Some examples of current members of the association illustrate the wide range of organizations represented by RIA: FedEx (user), Productive Robotics (supplier), PINC (supplier), Wichita State University (educator/researcher), PlastiComp (consultant), York Exponential (integrator), Boston University Department of Economics (educator/researcher), Magna Engineered Glass (user), and Edge Case Research (consultant). Membership categories consist of one to three levels: platinum, gold, and silver, depending on the category involved. Membership fees also depend on category, ranging from $150 (user, silver level) to $10,000 (supplier, platinum level).

RIA has been very successful in achieving its objectives. According to the most recent figures available, RIA members produced 34,606 robots valued at $1.9 billion in 2016. These numbers were up about 10 percent over 2015. By far the largest user of robots was the automotive industry, which accounted for 30,875 robots valued at $1.8 billion. Two of the largest industrial applications for robots were in assembly applications and spot welding. One of the largest growth rates for the year was in the food and consumer goods industry, where sales increased by 32 percent. Major uses of robots in the industry were improvement of food safety practices; performing repetitive primary packaging tasks such as bin picking, tray loading, and bottle handling; and assisting with secondary packaging tasks such as case packing, bundling, bagging, and pallet loading.

RIA's work consists of a number of activities, among the most important being educational events, such as webinars and conference. Some topics of recent meetings are "Collaborative Robots and Advanced Vision," "Automata 2019," and "Robot Safety Standard and Robot Risk Assessment Training." The association is also intimately involved in the development and promulgation of standards for the industry. Areas for which standards have been developed are vocabulary and characteristics, personal care robot safety, industrial safety, service robots, medical robot safety, and modularity for service robots.

RIA has a number of publications of interest and value both to members and, in some cases, to nonmembers. The Robotics Online Electronic Newsletter, for example, provides case studies, technical articles, and feature articles; trends in robotics and recent statistics; expert advice on specific problems in the field; and recent robotic products and applications. Another valuable resource, available for free download, is Collaborative Robots Whitepaper, laying out information about "the hottest area of interest" in the robotics interest. The paper answers questions such as what industries can make use of collaborative robots, what comparisons can be made between collaborative and traditional robots, and whether collaborative robots are a good choice for one's specific business.

For users in the field, the RIA Web site also has an ROI Robot System Value Calculator, which allows a user to calculate the savings available for replacing a human worker (or workers) with a robot over the lifetime of the device. The Web site's "Robotics Industries Statistics" is also a valuable source of information about the number and type of devices sold and used in a particular year, along with trends in the industry. Another section of the Web site is devoted to safety issues. It includes copies of current international safety standards for robots, along with RIA technical reports on the topic. It also lists safety training sessions available on the topic as well as in-house safety training opportunities. The RIA blog consists of articles on technical aspects of robot production and user, such as "Robotic Cable Management Best Practices," "Collaborative Robots and Cybersecurity Concerns," "Robot Safety Standards for Industrial Mobile Robots," and "Medical Robots: Top 4 Applications in Use Today."

Robotics Institute at Carnegie Mellon University

The Robotics Institute at Carnegie Mellon University was established in 1979, when two Carnegie Mellon professors, Raj Reddy and Angel Jordan, joined with the president of the Westinghouse Electric Corporation, Tom Murrin, to create a workspace that they hoped to be "the best place on the planet to do robotics research." The organization's record of achievement that time is impressive. It includes such developments as a one-legged, hopping robot that eventually became the prototype of such modern devices as Bow Leg Hopper and Thumper; the Remote Reconnaissance Vehicle, which was used for surveying and cleaning the floor of the damaged Three Mile Island nuclear reactor; an eight-legged robot used to collect samples of materials from active volcanoes; a robot designed to assist surgeons in hip replacement surgery; a social robot, Grace, capable of interacting with humans; and the first full-size helicopter capable of operating autonomously.

The institute maintains two major research centers, the National Robotics Engineering Center and the Field Robotics Center. As their names suggest, the former conducts basic research on the development of robotic devices, while the latter conducts field studies of such robots. Some current projects conducted at the institute include an adaptive traffic light signal, assistive robots for blind travelers, autonomous driving vehicles and devices, autonomous vehicle safety verification systems, an autonomous vineyard canopy and yield estimation system, biodegradable electronics, blood-plasma-based plastics, autonomous systems for specialty crops, and robotic watercraft.

Through its affiliation with Carnegie Mellon, the institute offers a variety of degree programs, including majors and minors in robotics, and MS degrees in robotics, robotic systems development, and computer vision, along with a PhD in robotics. In addition to formal course work, the institute offers a wide variety of other learning experiences, such as seminars, presentations of thesis proposal and thesis defenses, faculty talks, and social get-togethers.

Victor Scheinman (1942–2016)

Scheinman was a pioneer in the modern-day field of robotics. He is probably best remembered for his invention of the Stanford Arm, a six-jointed device that was capable of picking up and maneuvering objects, such as blocks, that the device was then able to arrange in some predetermined order. The arm located the necessary building blocks by means of a camera, which transmitted this information to a computer. The computer program then directed the arm as to the next steps to be performed in order to complete the desired structure. The arm was a modification of earlier robotic devices, beginning with the Rancho Arm, invented in the early 1960s by researchers at the Rancho Los Amigos Hospital in Downey, California. The arm was designed to assist individuals whose arms had become damaged. It was sold to Stanford University in 1963, where it underwent further development, one of which was known as the Stanford Hydraulic Arm. Scheinman's invention was the final step in the evolution of this early robotic device.

Victor David Scheinman was born in Augusta, Georgia, on December 28, 1942, shortly after his army officer father had been transferred to a base near Augustua. At the war's conclusion, the Scheinman family moved to Brooklyn where they remained for the next decade. Scheinman has said in an interview with the Robotics History project that his first experience with robotics came when he went to see a motion picture *The Day the Earth Stood Still*. The film frightened him so much, he said, that he had terrible nightmares about the robot for some time afterward. Over time, however, his exposure to robots became a permanent feature of his long-term interest in the field.

Scheinman attended New Lincoln High School at 31 West 110th Street in Manhattan. The school was an experimental endeavor set up by a group of parents whose children had previously attended the city's Horace Mann High School. One of Scheinman's projects during high school was the invention of a machine that converted spoken word to written text. After graduation from New Lincoln at the age of 16, Scheinman enrolled at the Massachusetts Institute of Technology (MIT), planning to major in aeronautics and astronautics, a field in which he earned his bachelor's degree in 1963. He seemed still on track to working in the field of aeronautics, having spent the summer after his junior year at the Sikorsky helicopter plant in Stratford, Connecticut.

Scheinman's postgraduate years were strongly influenced by an MIT advisor, who suggested that he spend some time in industry and then go to Stanford for his graduate studies. The advisor also arranged a job for him at Boeing company, where he remained for only four months. He then took the rest of the year off to "see the world," which meant traveling to Tahiti, New Zealand, Australia, Philippines, Thailand, and Malaysia. At the conclusion of that adventure, he returned to the United States and enrolled in a master's program in engineering at Stanford. It was during his year at Stanford that Scheinman invented the Stanford Arm. His project, he later explained, was one of taking a device (the Rancho Arm) which was based on a useful concept, but that still had too many technical problems associated with it.

On graduation from Stanford, Scheinman founded his own company for the production and sale of the Stanford Arm. He called the company Vicarm, short for Victor Arm. The company was later sold to Unimation company, where it was incorporated in the first industrial robots then being built by the company. Scheinman was made general manager of the company's West Coast division, after which he was involved in the founding of a new company, Automatix, where his major contribution was the inclusion of vision capability in robotic devices. Over the next three decades, Scheinman continued to serve as a consultant to robotics companies and as visiting professor in the Department of Mechanical Engineering at Stanford. He died on September 20, 2016, in Petrolia, California, at the age of 73.

Alan Turing (1912–1954)

Turing is probably best known to the general public as the man who solved the "Enigma" puzzle during World War II. The term applies to a coding machine developed by German researchers in the 1920s for both commercial and military uses. The machine allowed the Germans to send and receive coded messages to forces located anywhere in the European region. Such messages to German submarines were responsible for the destruction of substantial numbers of Allied transport and battle ships, establishing a hegemony in the Atlantic Ocean that might conceivably have been the turning point in World War II. Working with a team of cryptographers at the Bletchley Park Government Code and Cypher School (GC&CS) just north of London, Turing was able to break the Enigma code and destroy the German coding program.

In the field of robotics, Turing is best remembered for his invention of a test designed to tell the difference between a human and a machine acting as a human, a test that has become known as the Turing test. The test is simple in concept. A human sits on one side of an opaque screen and poses questions to a person/robot on the opposite side of the screen. If the object to whom the questions are posed provides answers that are "reasonable" by any human standard, it is classified as a "human." The Turing test today, in one of a number of forms, is often recommended as a way of deciding if an unknown device is actually equivalent to a real human. If the device responds to various stimuli in the same way a human does, then the device (robot) can legitimately be regarded as a human. Such determinations are important as robotics engineers continue to develop more and more sophisticated robots. They lead to such questions as, if a robot passes a Turing test, is it then subject to laws, regulations, moral restrictions, rights and privileges, and other features that humans take for granted?

Alan Mathison Turing was born in the Maida Vale district of London on June 23, 1912. His father, Julius Mathison Turing, was a civil servant who spent much of his career in India, while his mother, Ethel Sara Stoney Turing, came from a distinguished family of scholars. Among her ancestors was the eminent Irish physicist George Johnstone Stoney, probably best known today for his naming of the fundamental particle of electricity, the electron. During Alan's childhood, his parents traveled to and from India on a number of occasions, leaving him and his older brother with retired army colonel Ward and his wife during their absence. The boys were not especially fond of the couple, of whom the colonel, it was said, was "as remote and gruff as God the Father."

Turing showed an interest in the natural world as early as age 3, when he buried a broken toy soldier in the garden, hoping to grow a mended replica. His mother also described an event when they went shopping together. On that occasion, Alan sat on the curb of the street collecting metal filings while his mother was in a store although, she later said, she had no idea what he wanted to do with them or even how he knew they would be there.

In the expectation that Alan would follow the normal educational process of upper-middle and upper-class boys of the time (preferably Oxford or Cambridge), he was enrolled at St. Michael's pre-preparatory school in 1918, where one of this teachers was later to write to Alan's parents that the boy "had genius." In 1922, Turing moved on to the next stage of his education, preparatory school at Hazelhurst, near Tunbridge Wells. There he continued to pursue his earlier interests in mathematics and science, following a course of self-education in areas (such as organic chemistry) that were not part of the usual Hazelhurst curriculum.

In 1926, Turing moved on to the Sherborne public school (in British parlance, the term refers to a private school equivalent to an American high school). There he received mixed reviews, with some teachers recognizing his extraordinary talents and with others complaining about his failure to follow school rules, reluctance to move to final conclusions with proper grounding in a topic, and following his own academic interests outside of the regular curriculum. During his years at Hazelhurst, he also met and developed a strong personal attachment to a fellow classmate, Christopher Morcom. Chris's death on February 13, 1930, of bovine tuberculosis was one of the most emotionally wrenching events in Turing's life.

In 1931, Turing was accepted as a student at King's College, Cambridge, where he eventually earned first-class honors in mathematics. He graduated in 1935, at which time he was elected a fellow at King's College. That honor was based largely on a paper he had written on a fundamental and unsolved problem in mathematics, the so-called central limit theory, "On Computable Numbers, with an Application to the *Entscheidungsproblem*." (King's authorities were not aware that the problem had earlier been solved by Finnish mathematician Jarl Waldemar Lindeberg.)

In the period between 1936 and 1938, Turing traveled to the United States, where he studied with the famous mathematician Alonzo Church at Princeton University. On his return to England, and with the threat of world war on the horizon, Turing joined the GC&CS, where he was assigned to work on the Enigma cipher. His remarkable success in that work was not recognized until the British government declassified documents from the period in 1974. By that time, Turing had been dead for 20 years.

At the conclusion of the war, Turing left GC&CS and joined a research team at the Computing Machine Laboratory at the Victoria University of Manchester. During his tenure at Manchester, he was arrested for having same-sex relationships with a young man in the city. He was found guilty and forced to take female hormones to "correct" his condition. During that process, he was found dead on June 7, 1954, in his home at Wilmslow, Cheshire. A partially eaten poisoned apple found next to his bed led police to conclude that he had committed suicide, although positive proof for that suspicion has never been found, nor is it likely to be.

Turing has been honored posthumously in a number of ways. A number of technical and mathematical procedures have been named in his honor, including Turing completeness, Turing degree, Turing patterns, Turing reduction, and Turing switch. In addition, the Association of Computing Machinery has named its highest honor for Turing, and the Computer Science Department of the University of Texas at Austin has instituted a program of Turing Scholars for promising students in the field of computer science. In 2013, almost 60 years after Turing's death, Queen Elizabeth II granted him a royal pardon for his conviction in 1952.

Jacques de Vaucanson (1709–1782)

De Vaucanson is sometimes credited with being the father of robotics because of sophisticated automata that he built. The first of these devices, built in 1737, was called the Flute Player. The device was about five feet tall and painted white to represent a marble statue. It operated by means of a series of bellows that blew air through the automaton's lips, producing a flute-like sound. A second automaton was called the Tambourine (or Tabor) Player. It was one of the most complicated of all automatons produced at the time since it was able to play one instrument with one hand (the flute) and a second instrument with the other (the tambourine). De Vaucanson's masterpiece, however, was the Duck, another complex device that was made of copper and lead and had over 400 moving parts. It was designed not only to provide amusement for observers but also to teach some basic principles of animal life. The duck was able to reach out and accept food from a person's hand, swallow that food, and eventually to defecate the digested food. It was also to produce a number of other duck-like actions, such as quacking, flapping its wings, and drinking water.

Jacques de Vaucanson was born in Grenoble, France, on February 24, 1709. He was the tenth son of a glovemaker whose family lived in poverty. De Vaucanson imagined that one way to escape that poverty was to find a profitable career, and he chose clockmaking for that purpose. Among his complex automata, he also built a number of complex clocks during his life. Originally born as Jacques Vaucanson, the particle "de" ("of") was added to his name as an honorific. When clockmaking proved to be an inadequate way of making a living, de Vaucanson was sent to a Jesuit school in Grenoble (today the Lycée Stendhal) to study for the priesthood. In 1725, he completed his training, took his orders, and joined Les Ordre des Minimes in Lyon.

There is a perhaps apocryphal story that de Vaucanson maintained his fascination with invention while serving as a priest and was given a workshop to continue his studies of the devices. It is said, however, that one of his superiors thought that such work was "profane," and he ordered the workshop to be destroyed. Sensing frustration at the resistance to his experimental work, de Vaucanson left the priesthood in 1728 and moved to Paris, where he continued his work on clocks and other mechanical devices. While in Paris, he also studied medicine and anatomy at the Jardin du Roi. These activities were made possible by the generous financial support of Parisian financier Samuel Bernard, said to be one of the wealthiest men of his time.

In 1731, Vaucanson left Paris and moved to Rouen, where he developed a friendship with two physicians, Claude-Nichols

Le Cat and François Quesnay. A common interest among the three for the workings of the human body inspired de Vaucanson to direct his research to the construction of mechanical devices that represented the way in which the body functions. From this research came the Flute Player, the Tambourine Player, and the Duck (supposedly a simpler animal to begin with than an actual human). The creation of such objects met a field of interest then sweeping through France and most of Europe: automatons that looked and performed like humans or other animals. They were the source both of education for the masses about scientific principles and of pure entertainment and amusement. De Vaucanson spent most of 1732 traveling around France, exhibiting his inventions.

In 1741 de Vaucanson's life took a new turn when he was appointed inspector of silk manufacture in France by Cardinal André-Hercule de Fleury, chief minister of Louis XV. At the time, the French silk industry, once the leading source of silk in the world, had fallen behind a more modern and efficient system of manufacture practiced in Great Britain. In his new position, de Vaucanson invented an automatic weaving machine to improve the efficiency of the French industry. Unfortunately, his scheme was never successful, partly for technical reasons (adequate materials for the machine were not available) and partly because of public resistance (mainly from workers who feared loss of their jobs). Over the next two decades, de Vaucanson continued to work on more complex automata, with the ultimate goal of building a human figure similar to the Duck. These efforts never came to fruition, however, again largely because of the complexity of the task and the lack of adequate materials to use in the device. He died in Paris on November 21, 1782, at the age of 73.

Legal documents about robotics are relatively sparse. This chapter contains selections from some of the laws and regulations about robots that have been adopted, along with portions of court cases bearing on robotic issues. Data about robotics is also scarce. Some of the most fundamental of that information is also provided here.

Data

Table 5.1 Estimated Yearly Shipments of Multipurpose Industrial Robots in Selected Countries (number of units)

Country	2014	2015	2016	2019*
Brazil	1,266	1,407	1,800	3,500
Rest of South America	321	283	400	1,200
North America	31,029	36,444	38,000	46,000
China	57,096	68,556	90,000	160,000
India	2,126	2,065	2,600	6,000
Japan	29,297	35,023	38,000	43,000

(continued)

Survival Research Laboratories opens an exhibition of killing machines and robots in New York in 2018. The solo exhibition includes eight "kinetic sculptures" created by SRL between 1986 and the present, and videos of past exhibitions when the robotic machines came to life. Artist Mark Pauline, the founder of SRL, uses his art pieces to convey a "visceral, humanized vision of machines" that serves as a type of socio-political satire. (Scott Lynch/Pacific Press/LightRocket via Getty Images)

Table 5.1 (*continued*)

Country	2014	2015	2016	2019*
Republic of Korea	24,721	38,285	40,000	46,000
Taiwan	6,912	7,200	9,000	13,000
Thailand	3,657	2,556	3,000	4,500
Other Asia/Australia	10,635	6,873	7,600	13,200
Central/Eastern Europe	4,643	5,976	7,550	11,300
France	2,944	20,051	20,105	21,000
Germany	20,051	20,105	21,000	25,000
Italy	6,215	6,657	7,200	9,000
Spain	2,312	3,766	4,100	5,100
United Kingdom	2,094	1,645	1,800	2,500
Other Europe	7,300	8,879	9,250	11,400
Africa, unspecified countries	7,524	4,635	5,000	8,000
Total	220,571	253,748	290,000	414,000

*Forecast.

Source: Executive Summary World Robotics 2016 Industrial Robots. 2016. Table 1, page 18. https://ifr.org/img/uploads/Executive_Summary_WR_Industrial_Robots_20161.pdf. Used with kind permission of the International Federation of Robotics.

Table 5.2 Estimated Worldwide Annual Supply of Industrial Robots

Year	Thousands of Units
2003	81
2004	97
2005	120
2006	112
2007	114
2008	113
2009	60
2010	121
2011	166
2012	159
2013	178
2014	221
2015	254

Source: Executive Summary World Robotics 2016 Industrial Robots. 2016. Page 11. https://ifr.org/img/uploads/Executive_Summary_WR_Industrial_Robots_20161.pdf. Used with kind permission of the International Federation of Robotics.

Table 5.3 Worldwide Estimated Operational Stock of Industrial Robots

Year	Thousands of Units
1973	3
1983	66
1990	454
1995	605
2000	750
2005	923
2010	1,059
2014	1,472
2015	1,632
2016*	1,824
2019*	2,589

*Forecast.

Source: Welcome to IFR Press Conference. 2016. International Federation of Robotics, 5. https://ifr.org/downloads/press/02_2016/Presentation_market_over viewWorld_Robotics_29_9_2016.pdf. Accessed on September 16, 2017. Used with kind permission of the International Federation of Robotics.

Table 5.4 Use of Robots by Various Industries

Industry	2014	2015	2016–2019*
Logistic**	12,650	19,000	175,000
Defense	11,000	11,200	74,800
Field***	5,760	6,440	34,600
Cleaning	280	600	11,700
Medical	240	1,325	8,150
Mobile platforms	530	710	7,500
Exoskeletons	275	370	6,550
Public relations	105	245	6,500
Inspection	260	275	3,610
Construction	560	570	2,800
All others	290	320	2,000

*Includes forecast.
**Robots for transporting materials.
***Primarily agricultural robots.

Source: Welcome to IFR Press Conference. 2016. International Federation of Robotics, 8, 10. https://ifr.org/downloads/press/02_2016/Presentation_12_Oct_ 2016__WR_Service_Robots.pdf. Accessed on September 16, 2017. Used with kind permission of the International Federation of Robotics.

Table 5.5   Distribution of Robots among Various Countries and Regions
(thousands of units)

Country/Region	2010	2011	2012	2013	2014	2015	2016*	2017*	2018*	2019*
China	15	23	23	37	57	69	90	110	130	160
South Korea	24	26	19	21	25	38	40	42	44	46
Japan	22	28	29	25	29	35	38	39	41	43
Western Europe	23	32	30	30	34	35	37	39	42	46
Eastern Europe	3	5	4	5	5	6	8	9	10	11

*Forecast.

Source: Welcome to IFR Press Conference. 2016. International Federation of Robotics, 9–14. https://ifr.org/downloads/press/02_2016/Presentation_12_Oct_2016_WR_Service_Robots.pdf. Accessed on September 16, 2017. Used with kind permission of the International Federation of Robotics.

Documents

Guidelines for Robotic Safety (1999)

The U.S. Occupational Safety and Health Administration (OSHA) began writing regulations for the safe use of robots in the country in 1987. The most recent of those regulations dates to 1999. The following excerpt includes some of the most basic information provided in a somewhat abbreviated form. Triple asterisks (***) indicate the omission of material for this excerpt.

I. Introduction

[The introduction provides a general definition of robots and explanation of the tasks they perform in industry.]

A. Accidents: Past Studies

1. Studies in Sweden and Japan indicate that many robot accidents do not occur under normal operating conditions but, instead during programming, program touch-up or refinement, maintenance, repair, testing, setup, or adjustment. During many of these operations the operator, programmer, or corrective maintenance

worker may temporarily be within the robot's working envelope where unintended operations could result in injuries.

2. Typical accidents have included the following:

- A robot's arm functioned erratically during a programming sequence and struck the operator.

- A materials handling robot operator entered a robot's work envelope during operations and was pinned between the back end of the robot and a safety pole.

- A fellow employee accidentally tripped the power switch while a maintenance worker was servicing an assembly robot. The robot's arm struck the maintenance worker's hand.

B. Robot Safeguarding

1. The proper selection of an effective robotic safeguarding system should be based upon a hazard analysis of the robot system's use, programming, and maintenance operations. Among the factors to be considered are the tasks a robot will be programmed to perform, start-up and command or programming procedures, environmental conditions, location and installation requirements, possible human errors, scheduled and unscheduled maintenance, possible robot and system malfunctions, normal mode of operation, and all personnel functions and duties.

2. An effective safeguarding system protects not only operators but also engineers, programmers, maintenance personnel, and any others who work on or with robot systems and could be exposed to hazards associated with a robot's operation. A combination of safeguarding methods may be used. Redundancy and backup systems are especially recommended, particularly if a robot or robot system is operating in hazardous conditions or handling hazardous materials. The safeguarding devices

employed should not themselves constitute or act as a hazard or curtail necessary vision or viewing by attending human operators.

II. Types and Classifications of Robots
***

III. Hazards
***

I. **Types of Accidents**. Robotic incidents can be grouped into four categories: a robotic arm or controlled tool causes the accident, places an individual in a risk circumstance, an accessory of the robot's mechanical parts fails, or the power supplies to the robot are uncontrolled.

   1. **Impact or Collision Accidents**. Unpredicted movements, component malfunctions, or unpredicted program changes related to the robot's arm or peripheral equipment can result in contact accidents.

   2. **Crushing and Trapping Accidents**. A worker's limb or other body part can be trapped between a robot's arm and other peripheral equipment, or the individual may be physically driven into and crushed by other peripheral equipment.

   3. **Mechanical Part Accidents**. The breakdown of the robot's drive components, tooling or end-effector, peripheral equipment, or its power source is a mechanical accident. The release of parts, failure of gripper mechanism, or the failure of end-effector power tools (e.g., grinding wheels, buffing wheels, deburring tools, power screwdrivers, and nut runners) are a few types of mechanical failures.

   4. **Other Accidents**. Other accidents can result from working with robots. Equipment that supplies robot

power and control represents potential electrical and pressurized fluid hazards. Ruptured hydraulic lines could create dangerous high-pressure cutting streams or whipping hose hazards. Environmental accidents from arc flash, metal spatter, dust, electromagnetic, or radio-frequency interference can also occur. In addition, equipment and power cables on the floor present tripping hazards.

II. **Sources of Hazards**. The expected hazards of machine to humans can be expected with several additional variations, as follows.

1. Human Errors. Inherent prior programming, interfacing activated peripheral equipment, or connecting live input-output sensors to the microprocessor or a peripheral can cause dangerous, unpredicted movement or action by the robot from human error. The incorrect activation of the "teach pendant" or control panel is a frequent human error. The greatest problem, however, is over familiarity with the robot's redundant motions so that an individual places himself in a hazardous position while programming the robot or performing maintenance on it.

2. Control Errors. Intrinsic faults within the control system of the robot, errors in software, electromagnetic interference, and radio frequency interference are control errors. In addition, these errors can occur due to faults in the hydraulic, pneumatic, or electrical subcontrols associated with the robot or robot system.

3. Unauthorized Access. Entry into a robot's safeguarded area is hazardous because the person involved may not be familiar with the safeguards in place or their activation status.

4. Mechanical Failures. Operating programs may not account for cumulative mechanical part failure, and faulty or unexpected operation may occur.

5. Environmental Sources. Electromagnetic or radio-frequency interference (transient signals) should be considered to exert an undesirable influence on robotic operation and increase the potential for injury to any person working in the area. Solutions to environmental hazards should be documented prior to equipment start-up.

6. Power Systems. Pneumatic, hydraulic, or electrical power sources that have malfunctioning control or transmission elements in the robot power system can disrupt electrical signals to the control and/or power-supply lines. Fire risks are increased by electrical overloads or by use of flammable hydraulic oil. Electrical shock and release of stored energy from accumulating devices also can be hazardous to personnel.

7. Improper Installation. The design, requirements, and layout of equipment, utilities, and facilities of a robot or robot system, if inadequately done, can lead to inherent hazards.

***

V. Control and Safeguarding Personnel

For the planning stage, installation, and subsequent operation of a robot or robot system, one should consider the following.
[Detailed discussion of each item omitted.]

1. **Risk Assessment**. At each stage of development of the robot and robot system a risk assessment should be performed. ***

2. **Safeguarding Devices**. Personnel should be safeguarded from hazards associated with the restricted envelope (space) through the use of one or more safeguarding devices:

   • Mechanical limiting devices;
   • Nonmechanical limiting devices;

- Presence-sensing safeguarding devices;
- Fixed barriers (which prevent contact with moving parts); and
- Interlocked barrier guards.

3. **Awareness Devices.** Typical awareness devices include chain or rope barriers with supporting stanchions or flashing lights, signs, whistles, and horns. ***

4. **Safeguarding the Teacher.** Special consideration must be given to the teacher or person who is programming the robot. ***

5. **Operator Safeguards.** The system operator should be protected from all hazards during operations performed by the robot. ***

6. **Attended Continuous Operation.** When a person is permitted to be in or near the robots restricted envelope to evaluate or check the robots motion or other operations, all continuous operation safeguards must be in force. During this operation, the robot should be at slow speed, and the operator would have the robot in the teach mode and be fully in control of all operations.

7. **Maintenance and Repair Personnel.** Safeguarding maintenance and repair personnel is very difficult because their job functions are so varied. Troubleshooting faults or problems with the robot, controller, tooling, or other associated equipment is just part of their job. Program touchup is another of their jobs as is scheduled maintenance, and adjustments of tooling, gages, recalibration, and many other types of functions. ***

8. **Maintenance.** Maintenance should occur during the regular and periodic inspection program for a robot or robot system. An inspection program should include, but not be limited to, the recommendations of the robot manufacturer and manufacturer of other associated robot system equipment such as conveyor mechanisms, parts feeders, tooling, gages, sensors, and the like. ***

9. **Safety Training**. Personnel who program, operate, maintain, or repair robots or robot systems should receive adequate safety training, and they should be able to demonstrate their competence to perform their jobs safely. ***

10. **General Requirements**. To ensure minimum safe operating practices and safeguards for robots and robot systems covered by this instruction, the following sections of the ANSI/RIA R15.06-1992 must also be considered. ***

Source: Section IV: Chapter 4. n.d. *ODHA Technical Manual.* https://www.osha.gov/dts/osta/otm/otm_iv/otm_iv_4.html. Accessed on September 17, 2017.

Efficacy of Robotic Surgery (2013)

One of the most basic questions about the use of robots for surgical procedures is whether or not they are safe for general use. In 2013, the FDA conducted a "small sample survey" to obtain feedback from surgeons who had actually used the da Vinci surgical system in their own practices. The conclusions the agency drew from this study are as follows. Triple asterisks (***) indicate omitted text.

Summary
***

All respondents report that learning how to use the da Vinci Surgical System is the biggest challenge because of the device's complex user-interface. All respondents have participated in some type of training, whether it was in their hospital or at one of the manufacturer's training centers. All respondents say they needed time for learning how to use the foot pedals, acquiring effective hand-eye coordination, and performing procedures without the ability to use their hands to touch or feel tissues, organs, or use sutures. For cardiothoracic surgeons, it may be more of a challenge learning to use the da Vinci Surgical

System, since they may not have any prior laparoscopic training and experience. In terms of proficiency, respondents vary in the number of procedures they performed before feeling comfortable using the device. Seven of 11 respondents report proctoring other surgeons but only those in their own hospitals.

All respondents indicate that selecting the appropriate patient for surgery using the da Vinci Surgical System is a determinant for successful patient outcomes. All respondents report fewer patient complications and shorter hospital stays as a benefit of surgery with the system. Respondents who had to convert to an open procedure did so more often because of the patient's anatomy or co-morbidities, rather than because of system problems.

Ten of the 11 respondents report problems with the arms of the da Vinci Surgical System. Some of the arms were repaired by the manufacturer; others were replaced. Of note, only three of the 11 respondents are aware of any recalls with the da Vinci Surgical System.

Respondents' suggestions for enhancing the da Vinci Surgical System range from changes in the system design and training, to mounting the device on the ceiling to allow more space in smaller rooms.

Respondents offer the following suggestions:

- More training hours in the dry lab;
- Mandated hours for simulation training;
- Haptic or tactile feedback to determine the condition of the tissues and how hard to pull or push on tissues and sutures;
- Integration of ultrasound into the program for kidney procedures; ultrasound is currently controlled by the first assistant. According to one respondent, the manufacturer is developing a kidney ultrasound.
- Smaller arms;
- Smaller instruments;

- Integration of all information surgeons need when performing procedures, i.e., patient data including pulse rate, blood loss, oxygen saturation, etc. Staff in the operating room suite sees the data, except the surgeon. A button or pedal to push to see patient data as needed would be helpful; and,

- A clock would be helpful.

Source: Small Sample Survey—Final Report. 2013. FDA. https://www.fda.gov/downloads/MedicalDevices/Products andMedicalProcedures/SurgeryandLifeSupport/ComputerAssi stedSurgicalSystems/UCM374095.pdf. Accessed on September 19, 2017.

Idaho Drone Law (2013)

Drones are a common type of robot with which the average person might come into contact. It appears that drones will be much more commonly used in the future, as for the delivery of package of mail order companies. Thus far, relatively few laws have been passed to deal with the possible risks posed by drones. The Idaho law cited here is one of the first of such laws.

21-213. RESTRICTIONS ON USE OF UNMANNED AIRCRAFT SYSTEMS—DEFINITION—VIOLATION—CAUSE OF ACTION AND DAMAGES.

(1)(a) For the purposes of this section, the term "unmanned aircraft system" (UAS) means an unmanned aircraft vehicle, drone, remotely piloted vehicle, remotely piloted aircraft or remotely operated aircraft that is a powered aerial vehicle that does not carry a human operator, can fly autonomously or remotely and can be expendable or recoverable.

(b) Unmanned aircraft system does not include:

(i) Model flying airplanes or rockets including, but not necessarily limited to, those that are radio controlled or otherwise remotely controlled and that are used purely for sport or recreational purposes; and

(ii) An unmanned aircraft system used in mapping or resource management.

(2)(a) Absent a warrant, and except for emergency response for safety, search and rescue or controlled substance investigations, no person, entity or state agency shall use an unmanned aircraft system to intentionally conduct surveillance of, gather evidence or collect information about, or photographically or electronically record specifically targeted persons or specifically targeted private property including, but not limited to:

   (i) An individual or a dwelling owned by an individual and such dwelling's curtilage, without such individual's written consent;

   (ii) A farm, dairy, ranch or other agricultural industry without the written consent of the owner of such farm, dairy, ranch or other agricultural industry.

 (b) No person, entity or state agency shall use an unmanned aircraft system to photograph or otherwise record an individual, without such individual's written consent, for the purpose of publishing or otherwise publicly disseminating such photograph or recording.

(3) Any person who is the subject of prohibited conduct under subsection (2) of this section shall:

 (a) Have a civil cause of action against the person, entity or state agency for such prohibited conduct; and

 (b) Be entitled to recover from any such person, entity or state agency damages in the amount of the greater of one thousand dollars ($1,000) or actual and general damages, plus reasonable attorney's fees and other litigation costs reasonably incurred.

(4) An owner of facilities located on lands owned by another under a valid easement, permit, license or other right of

occupancy is not prohibited in this section from using an unmanned aircraft system to aerially inspect such facilities.

Source: Title 21: Aeronautics. 2013. Idaho Statutes. https:// legislature.idaho.gov/statutesrules/idstat/title21/t21ch2/sect21-213/. Accessed on September 18, 2017.

Robotically Assisted Surgery (2015)

Robotically assisted surgery (RAS) is a fairly new field of medicine. Products and applications in the field must receive the approval of the U.S. Food and Drug Administration (FDA) before they can be used with the general public. In 2015, the FDA published a general summary of the work involved in its regulatory process, along with information for practitioners in the field and the general public. The following is an excerpt from this report.

What Are Computer-Assisted Surgical Systems?

Different types of computer-assisted surgical systems can be used for pre-operative planning, surgical navigation and to assist in performing surgical procedures. Robotically-assisted surgical (RAS) devices are one type of computer-assisted surgical system. Sometimes referred to as robotic surgery, RAS devices enable the surgeon to use computer and software technology to control and move surgical instruments through one or more tiny incisions in the patient's body (minimally invasive) for a variety of surgical procedures.

The benefits of a RAS device may include its ability to facilitate minimally invasive surgery and assist with complex tasks in confined areas of the body. The device is not actually a robot because it cannot perform surgery without direct human control.

RAS devices generally have several components, which may include:

- A console, where the surgeon sits during surgery. The console is the control center of the device and allows the surgeon to view the surgical field through a 3D endoscope and control movement of the surgical instruments;
- The bedside cart that includes three or four hinged mechanical arms, camera (endoscope) and surgical instruments that the surgeon controls during surgical procedures; and
- A separate cart that contains supporting hardware and software components, such as an electrosurgical unit (ESU), suction/irrigation pumps, and light source for the endoscope.

Most surgeons use multiple surgical instruments and accessories with the RAS device, such as scalpels, forceps, graspers, dissectors, cautery, scissors, retractors and suction irrigators.

Common Uses of Robotically-Assisted Surgical (RAS) Devices
[This section omitted.]

Recommendations for Patients and Health Care Providers about Robotically-Assisted Surgery

Health Care Providers

Robotically-assisted surgery is an important treatment option that is safe and effective when used appropriately and with proper training. The FDA does not regulate the practice of medicine and therefore does not supervise or provide accreditation for physician training nor does it oversee training and education related to legally marketed medical devices. Instead, training development and implementation is the responsibility of the manufacturer, physicians, and health care facilities. In some cases, professional societies and specialty board certification organizations may also develop and support training for their specialty physicians. Specialty boards also maintain certification status of their specialty physicians.

Physicians, hospitals and facilities that use RAS devices should ensure that proper training is completed and that surgeons have appropriate credentials to perform surgical procedures with these devices. Device users should ensure they maintain their credentialing. Hospitals and facilities should also ensure that other surgical staff that use these devices complete proper training.

Users of the device should realize that there are several different models of robotically-assisted surgical devices. Each model may operate differently and may not have the same functions. Users should know the differences between the models and make sure to get appropriate training on each model.

If you suspect a problem or complications associated with the use of RAS devices, the FDA encourages you to file a voluntary report through MedWatch, the FDA Safety Information and Adverse Event Reporting program. Health care personnel employed by facilities that are subject to FDA's user facility reporting requirements should follow the reporting procedures established by their facilities. Prompt reporting of adverse events can help the FDA identify and better understand the risks associated with medical devices.

Patients

Robotically-assisted surgery is an important treatment option but may not be appropriate in all situations. Talk to your physician about the risks and benefits of robotically-assisted surgeries, as well as the risks and benefits of other treatment options.

Patients who are considering treatment with robotically-assisted surgeries should discuss the options for these devices with their health care provider, and feel free to inquire about their surgeon's training and experience with these devices.

Source: Computer-Assisted Surgical Systems. 2015. FDA. https://www.fda.gov/MedicalDevices/ProductsandMedicalPro

cedures/SurgeryandLifeSupport/ComputerAssistedSurgical
Systems/default.htm. Accessed on September 19, 2017.

Florida Law on Driverless Vehicles (2016)

The development of driverless (autonomous) vehicles has led to some critical issues in traffic safety. Such vehicles are no longer the dreams of automotive designers and engineers but are recording thousands of miles in real traffic in every part of the United States. So far, no federal regulations cover autonomous vehicles specifically, but a number of states have taken up the problem. In 2011, Nevada became the first state to adopt legislation covering the use of autonomous vehicles. The new law turned out to be unworkable, however, and was revised two years later and rewritten in its current form in 2017. Various states take different approaches to the regulation of autonomous vehicles, ranging from requirements for state testing on driverless vehicles to full approval of their licensing in the state (see "Autonomous Vehicles," http://www.ncsl.org/ research/transportation/autonomous-vehicles-self-driving-vehicles-enacted-legislation.aspx). An example of state with the latter of these options is the Florida law, from which pertinent sections are reprinted here. Triple asterisks (***) represent the omission of text from the law.

Section 7. Subsection (1) of section 316.85, Florida Statutes, is amended to read:
  316.85 Autonomous vehicles; operation.—

(1) A person who possesses a valid driver license may operate an autonomous vehicle in autonomous mode on roads in this state if the vehicle is equipped with autonomous technology, as defined in s. 316.003.

***

Section 8. Section 316.86, Florida Statutes, is amended to read:
  316.86 Exemption from liability for manufacturer when third party converts vehicle.—

The original manufacturer of a vehicle converted by a third party into an autonomous vehicle is not liable in, and shall have a defense to and be dismissed from, any legal action brought against the original manufacturer by any person injured due to an alleged vehicle defect caused by the conversion of the vehicle, or by equipment installed by the converter, unless the alleged defect was present in the vehicle as originally manufactured.

Section 9. Subsection (1) of section 319.145, Florida Statutes, is amended to read:

319.145 Autonomous vehicles.—

(1) An autonomous vehicle registered in this state must continue to meet applicable federal standards and regulations for such a motor vehicle. The vehicle must:

(a) Have a system to safely alert the operator if an autonomous technology failure is detected while the autonomous technology is engaged. When an alert is given, the system must:

1. Require the operator to take control of the autonomous vehicle; or

2. If the operator does not, or is not able to, take control of the autonomous vehicle, be capable of bringing the vehicle to a complete stop.

(b) Have a means, inside the vehicle, to visually indicate when the vehicle is operating in autonomous mode.

(c) Be capable of being operated in compliance with the applicable traffic and motor vehicle laws of this state.

Source: HB 7027, Engrossed 1. 2016. Florida House of Representatives. http://www.myfloridahouse.gov/Sections/Documents/loaddoc.aspx?FileName=_h7027er.docx&DocumentType=Bill&BillNumber=7027&Session=2016. Accessed on September 18, 2017.

Draft Report (on Robotics) of the European Parliament (2016)

The European Union has given substantial attention to the issue of robotics in today's world and in the future. It has considered a range of questions, such as the legal status of robots, their applications in a variety of occupations, and possible consequences of their interactions with humans. In 2016, the European Parliament's Committee on Legal Affairs released a Draft Report containing its recommendations with regard to the steps that might be taken by the Commission on Civil Law Rules on Robotics to deal with these issues. Some extracts of that report are reprinted here. Triple asterisks (***) represent the omission of text.

[Initial sections of the report deal with an introduction to robot technology and its likely applications in the future, general principles, and liability. Then follows:]

General Principles Concerning the Development of Robotics and Artificial Intelligence for Civil Use

1. Calls on the Commission to propose a common European definition of smart autonomous robots and their subcategories by taking into consideration the following characteristics of a smart robot:

   - acquires autonomy through sensors and/or by exchanging data with its environment (inter-connectivity) and trades and analyses data
   - is self-learning (optional criterion)
   - has a physical support
   - adapts its behaviours and actions to its environment;

2. Considers that a system of registration of advanced robots should be introduced, and calls on the Commission to establish criteria for the classification of robots with a view to identifying the robots that would need to be registered;

3. Underlines that many robotic applications are still in an experimental phase; welcomes the fact that more and more research projects are being funded with national and European money; calls on the Commission and the Member States to strengthen financial instruments for research projects in robotics and ICT; emphasises that sufficient resources need to be devoted to the search for solutions to the social and ethical challenges that the technological development and its applications raise;

4. Asks the Commission to foster research programmes that include a mechanism for short-term verification of the outcomes in order to understand what real risks and opportunities are associated with the dissemination of these technologies; calls on the Commission to combine all its effort in order to guarantee a smoother transition for these technologies from research to commercialisation on the market;

Ethical Principles

5 Notes that the potential for empowerment through the use of robotics is nuanced by a set of tensions or risks relating to human safety, privacy, integrity, dignity, autonomy and data ownership;

6. Considers that a guiding ethical framework for the design, production and use of robots is needed to complement the legal recommendations of the report and the existing national and Union acquis; proposes, in the annex to the resolution, a framework in the form of a charter consisting of a code of conduct for robotics engineers, of a code for research ethics committees when reviewing robotics protocols and of model licences for designers and users;

7. Points out that the guiding ethical framework should be based on the principles of beneficence, non-maleficence

and autonomy, as well as on the principles enshrined in the EU Charter of Fundamental Rights, such as human dignity and human rights, equality, justice and equity, non-discrimination and non-stigmatisation, autonomy and individual responsibility, informed consent, privacy and social responsibility, and on existing ethical practices and codes;

***

Intellectual Property Rights and the Flow of Data

10. Notes that there are no legal provisions that specifically apply to robotics, but that existing legal regimes and doctrines can be readily applied to robotics while some aspects appear to need specific consideration; calls on the Commission to come forward with a balanced approach to intellectual property rights when applied to hardware and software standards, and codes that protect innovation and at the same time foster innovation; calls on the Commission to elaborate criteria for an "own intellectual creation" for copyrightable works produced by computers or robots;

11. Calls on the Commission and the Member States to ensure that, in the development of any EU policy on robotics, privacy and data protection guarantees are embedded in line with the principles of necessity and proportionality; calls, in this regard, on the Commission to foster the development of standards for the concepts of privacy by design and privacy by default, informed consent and encryption;

12. Points out that the use of personal data as a "currency" with which services can be "bought" raises new issues in need of clarification; stresses that the use of personal data as a "currency" must not lead to a circumvention of the basic principles governing the right to privacy and data protection;

Standardisation, Safety and Security

13. Calls on the Commission to continue to work on the international harmonisation of technical standards, in particular together with the European Standardisation Organisations and the International Standardisation Organisation, in order to avoid fragmentation of the internal market and to meet consumers' concerns; asks the Commission to analyse existing European legislation with a view to checking the need for adaption in light of the development of robotics and artificial intelligence;

14. Emphasises that testing robots in real-life scenarios is essential for the identification and assessment of the risks they might entail, as well as of their technological development beyond a pure experimental laboratory phase; underlines, in this regard, that testing of robots in real-life scenarios, in particular in cities and on roads, raises numerous problems and requires an effective monitoring mechanism; calls on the Commission to draw up uniform criteria across all Member States which individual Member States should use in order to identify areas where experiments with robots are permitted;

[The report then concludes with recommendations for the use of robots in autonomous vehicles, care robots, medical robots, human repair and enhancement, and drones, as well as ancillary issues such as education and employment, liability, and international aspects of the issue.]

Source: Draft Report. 2016. Committee on Legal Affairs. European Parliament. http://www.europarl.europa.eu/sides/getDoc.do?pubRef=-//EP//NONSGML+COMPARL+PE-582.443+01+DOC+PDF+V0//EN&language=EN. Accessed on September 18, 2017.

National Robotics Initiative (2011/2016)

In June 2011, President Barack Obama announced the launch of the Advanced Marketing Partnership, a program designed to reinvigorate the field of manufacturing in the United States. One aspect of that program was the creation of the National Robotics Initiative, a program providing for grants for research on "developing robots that work with or beside people to extend or augment human capabilities, taking advantage of the different strengths of humans and robots." The program has been updated on a regular basis, with each year bringing some new perspectives on and opportunities in research on robotics. The most recent revision of the program, called National Robotics Initiative 2.0, was announced in 2016. The following excerpt is taken from NRI 2.0, but readers may also be interested in earlier versions of the program, all of which can be accessed from the source document listed at the end of this entry. Excerpts have been edited for brevity.

Synopsis of Program

The goal of the National Robotics Initiative (NRI) is to support fundamental research that will accelerate the development and use of robots in the United States that work beside or cooperatively with people. The original NRI program focused on innovative robotics research that emphasized the realization of collaborative robots (co-robots) working in symbiotic relationships with human partners. The NRI-2.0 program significantly extends this theme to focus on issues of scalability: how teams of multiple robots and multiple humans can interact and collaborate effectively; how robots can be designed to facilitate achievement of a variety of tasks in a variety of environments, with minimal modification to the hardware and software; how robots can learn to perform more effectively and efficiently, using large pools of information from the cloud, other robots, and other people; and how the design of the robots' hardware and software can facilitate large-scale, reliable operation. In addition, the program supports innovative approaches to establish and infuse robotics into educational curricula, advance the robotics workforce through education pathways, and explore the social, behavioral, and economic implications of our future with ubiquitous collaborative robots. Collaboration between academic, industry, non-profit, and other organizations is encouraged to establish better linkages between fundamental science and engineering and technology development, deployment and use. Well-justified international collaborations that add significant value to the proposed research and education activities will also be considered.

The NRI-2.0 program is supported by multiple agencies of the federal government including the National Science Foundation (NSF), the U.S. Department of Agriculture (USDA), the U.S. Department of Energy (DOE), and the U.S. Department of Defense (DOD). Questions concerning a particular project's focus, direction and relevance to a participating funding organization should be addressed to that agency's point of contact listed in section VIII of this solicitation.

***

I. Introduction

Robots—smart electro-mechanical devices that sense and operate within the environment of their surroundings—have the potential to transform our lives for the better. Specialized collaborative robots (co-robots) will safely assist people in their work and daily activities, while other robots will perform jobs too dangerous for people. We envision a future in which co-robots will no longer be expensive novelties, but rather ubiquitous technologies that significantly enrich the quality of life and quality of work for each of us.

Building upon foundational research in co-robots that began with the National Robotics Initiative (NRI), in NRI-2.0 NSF seeks research to help achieve this vision of ubiquitous collaborative robots, where robots are as commonplace as today's automobiles, computers, and cell phones. Robots will be found in homes, offices, hospitals, factories, farms, and mines; and in the air, on land, under water, and in space. Robots will be helping the elderly and people with disabilities in their activities of daily living. Robots will assist workers on the factory floor, performing mundane or dangerous tasks and helping to monitor worker safety. Robots will be among the first responders at natural disasters. We envision teams of humans and co-robots, large and small, reliably and efficiently cooperating on tasks. We envision democratizing robotics and transforming industries, benefiting the individual and society.

The NRI-2.0 program seeks research on the fundamental science, approaches, technologies, and integrated systems needed to achieve this vision. The program extends and advances the co-robot theme of the first five years of the NRI, in terms of the scale and variety of collaborative interactions. While the previous NRI co-robot theme focused on single robots collaborating with single humans, the NRI-2.0 program expands that theme to focus on issues relating to scaling up the technologies in ways that are necessary to achieve the vision of ubiquitous collaborative robots. One significant new aspect of this is co-robot teams: multiple co-robots collaborating with multiple people. Such teams will need to coordinate with people and robots, and possibly interact with software agents or make use of other devices (such as cell phones or the Internet of Things). Robots are characterized by embodied intelligence, and fundamental advances are needed in both the physical and digital domains. Furthermore, the added dimension of human interaction and the requirement to safely and productively work with human partners requires fundamental advances in modeling of human perception and cognition. To achieve the overall vision of this program, co-robot teams will facilitate communication not only through transfer of data, but also through physical and emotional channels. Collaboration will be enabled by innovative sensing and actuation schemes, and by new ways to leverage resources from the cloud. Co-robot teams will anticipate the behavior and needs of others, plan and learn from both human and robot collaborators, reason about alternate strategies, and allocate resource usage to promote efficient and effective collaboration.

While autonomy, embodiment, and human-machine interfaces have been extensively investigated separately, creating effective corobot teams will require understanding of the interactions between these elements. In particular, the physical embodiment of ubiquitous co-robots will substantially affect both how they perform and how they are perceived. Research is needed to understand, and take advantage of, the fundamental differences between embodied systems, such as robots, and virtual agents.

To scale up effectively, robots will need to be easily customizable and personalizable. Features of both the hardware and software should facilitate robots achieving a wide variety of tasks, in a wide variety of situations, for a wide diversity of people. Ubiquitous corobots should be designed to operate reliably and safely in real-world environments, with significant mean time between failure, in unstructured, uncertain environments. They should be designed for maintenance and incorporate capabilities for self-diagnosis and self-repair. Scaling up will require new approaches to the challenges of accountability, interoperability, and trust. Scalability also implies lowering the barrier to entry in robotics research, in terms of accessible, composable hardware and software infrastructure, and shareable testbeds and other resources that can be easily accessed and used by the larger research community.

Finally, the advent of ubiquitous co-robots will bring to the forefront social, economic, ethical, and legal issues. This program encourages fundamental research on the social and economic impact of robots on our work, our social institutions, and our quality of life and work. Pertinent research questions include understanding the complexities of the future co-robot economy; how economic and social inequality will be affected by ubiquitous co-robots; and what societal policies can be instituted to ensure that stakeholder groups can benefit from the presence of co-robots in our everyday lives.

***

[Each participating agency has some specific topics and areas in which they are looking for research. Some of the proposed topics suggested are as follows:]

Department of Defense

(1) Investigating socially-designed cues such as humanoid appearance, voice, personality, and other social elements on human trust and overall human-robot team performance; (2) physical "embodiment" features versus non-physical features to determine which have the most influence on human trust and performance; (3) sensing of human intent, cognitive and affective states, such as workload, stress, fatigue and fear; (4) modeling the processes of high-performing human teams, such as teammate monitoring, backup behavior, joint attention, shared mental models, coordination and negotiation; (5) dynamic modeling of the human-robot partnerships to allow continuous improvement of joint performance in real-world applications; (6) investigations regarding the effectiveness of various models of human-robot interaction, such as delegation and supervisory control; (7) practical methods for robotic systems to sense and measure trust and changes in trust over time; and (8) investigations of the impact of culture and cross-cultural interactions on reliance and human-machine cooperation.

Department of Energy

Wearable robotic devices for workers; traditional PPE *[personal protective equipment]* (e.g., hard hats, safety glasses, and steel-toed shoes); gaining remote access; glovebox operations; Multi-use and Multi-user (MU2) robotic technologies.

Department of Agriculture

Automated systems for planting, scouting, spraying, culturing, irrigating, and harvesting plant crops; improved robotics for inspection, monitoring, culturing, sorting, and handling of plants and flowers in controlled environment facilities and nurseries, or for managing or studying (e.g., monitoring, inspecting, sorting, vaccinating, deworming) large numbers of live animals, either domestic or wild; automated systems for inspection, sorting, processing, or handling of animal or plant products (including forest products) in post-harvest, processing, or product distribution environments; and multi-modal and rapid sensing systems for detecting defects, ripeness, physical damage, microbial contamination, size, shape, and other quality attributes of plant or animal products (including forest products), or for monitoring air or water quality.

Department of Education
(Examples of activities):

Design of innovative robotic technologies as tools for enhancing STEM learning in formal and informal learning environments;

Applications that further the development of co-robot systems that support personalized learning;

Design, implementation, and rigorous study of robotics competitions that impact student engagement, motivation to learn STEM content, and STEM career motivation;

Research and development of learning experiences and instructional models that integrate co-robotics within STEM courses;

Research of learning environments and instructional approaches in formal and informal settings to advance workforce preparedness in robotics; and Education research and development of strategies for broadening participation in education pathways to careers in robotics.

Source: National Robotics Initiative 2.0: Ubiquitous Collaborative Robots (NRI-2.0). 2016. National Science Foundation. https://www.nsf.gov/pubs/2017/nsf17518/nsf17518.pdf. Accessed on September 20, 2017.

Preparing for the Future of Artificial Intelligence (2016)

In October 2016, the Committee on Technology of the National Security and Technology Council released a paper on the future of artificial intelligence in the United States. The paper contained a number of mentions of robotic systems, one section of which is reproduced here. The section discusses "the sorts of practical problems that arise in making such a robot [a 'housecleaning robot'] effective and safe."

Avoiding Negative Side Effects: How can we ensure that our cleaning robot will not disturb the environment in negative ways while pursuing its goals, e.g., by knocking over a vase because it can clean faster by doing so? Can we do this without manually specifying everything the robot should not disturb?

Avoiding Reward Hacking: How can we ensure that the cleaning robot won't game its reward function? For example, if we reward the robot for achieving an environment free of messes, it might disable its vision so that it won't find any messes, or cover over messes with materials it can't see through, or simply hide when humans are around so they can't tell it about new types of messes.

Scalable Oversight: How can we efficiently ensure that the cleaning robot respects aspects of the objective that are too expensive to be frequently evaluated during training? For instance, it should throw out things that are unlikely to belong to anyone, but put aside things that might belong to someone (it should handle stray candy wrappers differently from stray cellphones). Asking the humans involved whether they lost anything can serve as a check on this, but this check might have to be relatively infrequent—can the robot find a way to do the right thing despite limited information?

Safe Exploration: How do we ensure that the cleaning robot doesn't make exploratory moves with very bad repercussions? For example, the robot should experiment with mopping strategies, but putting a wet mop in an electrical outlet is a very bad idea.

Robustness to Distributional Shift: How do we ensure that the cleaning robot recognizes, and behaves robustly, when in an environment different from its training environment? For example, heuristics it learned for cleaning factory work floors may be outright dangerous in an office.

Source: "Preparing for the Future of Artificial Intelligence." 2016. Committee on Technology. National Science and Technology Council. https://obamawhitehouse.archives.gov/sites/default/files/whitehouse_files/microsites/ostp/NSTC/preparing_for_the_future_of_ai.pdf. Accessed on September 17, 2017.

Regulatory Robot (CPSC) (2016)

Robots have permeated nearly every aspect of society today. As an example of a type of robot in an application that one might not otherwise expect, the following describes the U.S. Consumer Product Safety Commission's Regulatory Robot, an online device for helping businesses sort their way through regulatory procedures. The announcement of the robot's release provides basic information about the device. Triple asterisks (***) represent omitted text.

We are excited to introduce the Regulatory Robot, an important new tool and a step forward in the way that the CPSC works with small businesses. The Robot helps companies determine which consumer product safety rules might apply to their product. The Robot asks small businesses making new products a series of guided interview questions, and, based on the answers, produces a downloadable report. The report is customized with links to product safety regulations that may apply to the product. The report also provides important information on labeling, certification, and testing requirements.

The Robot, working in concert with the agency's other efforts to educate small businesses, will increase the overall level of consumer product safety in the United States.

***

Why Is This New Tool Noteworthy?

While information about CPSC rules and regulations has been available for years in a piecemeal fashion, this interactive tool is the first time it has been placed in proper context into one place on a government website.

The CPSC needs the cooperation and collaboration of small businesses to make sure that all consumer goods in the hands of American consumers are safe and compliant.

Why Did the CPSC Develop the Regulatory Robot?

Many small businesses seek assistance from our agency, through the Small Business Ombudsman and the Office of Compliance and Field Operations, and express a strong willingness to comply with our safety requirements—but they often cannot simply figure out how to do it. Prior to the Robot, we handled questions from small businesses through individualized phone calls to staff, an approach that challenged our limited resources. Small businesses will now have direct access to these requirements—and the Small Business Ombudsman and Office of Compliance and Field Operations will continue to be available for additional questions.

We believe that the Robot will lead to safer, compliant products, fewer deaths and injuries to American consumers, and help us further meet our mission. We believe that the Robot offers an excellent "safety" return on our investment, and we hope to see it evolve and grow further during the years with constructive user feedback.

Source: "New Regulatory Robot Tool Released: Enhanced Safety Guidance to Small Consumer Product Businesses on Compliance." CPSC. https://onsafety.cpsc.gov/blog/2016/01/07/ new-regulatory-robot-tool-released-enhanced-safety-guid ance-to-small-consumer-product-businesses-on-compliance/. Accessed on September 17, 2017.

Virginia Robot Delivery Laws (2017)

One of the most common ways in which individuals may be interacting with robots over the next few years will occur when a robot delivers a package directly to the steps of their home. Although technologically now possible, robot delivery poses a number of practical issues for merchants, customers, and governmental entities through which those robots will fly. In 2017, the Commonwealth of Virginia became the first state to pass legislation specifically addressing this issue. The relevant act amended Title 46.2, "Motor Vehicles," of the Code of Virginia. (The devices are referred to in the act as "electric personal delivery devices.") Triple asterisks (***) denote omitted text.

§ 46.2-100. Definitions

"Electric personal delivery device" means an electrically powered device that (i) is operated on sidewalks, shared-use paths, and crosswalks and intended primarily to transport property; (ii) weighs less than 50 pounds, excluding cargo; (iii) has a maximum speed of 10 miles per hour; and (iv) is equipped with technology to allow for operation of the device with or without the active control or monitoring of a natural person.

***

§ 46.2-908.1:1. Electric personal delivery devices.

A. All electric personal delivery devices shall obey all traffic and pedestrian control devices and signs and include a plate or marker that is in a position and size to be clearly visible and identifies the name and contact information of the owner of the electric personal delivery device and a unique identifying device number.

B. All electric personal delivery devices shall be equipped with a braking system that, when active or engaged, will enable such electric personal delivery device to come to a controlled stop.

C. No electric personal delivery device shall transport hazardous materials, substances, or waste as defined in § 10.1-1400.

For the purposes of this subsection, hazardous materials includes ammunition.

D. No electric personal delivery device shall be operated on a public highway in the Commonwealth, except to the extent necessary to cross an intersection or crosswalk.

E. No electric personal delivery device shall operate on a sidewalk or shared-use path or across a roadway on a crosswalk unless an electric personal delivery device operator is actively controlling or monitoring the navigation and operation of the electric personal delivery device.

F. Any entity or person who uses an electric personal delivery device to engage in criminal activity is criminally liable for such activity.

Source: 2017 Session. Chapter 788. LIS: Virginia's Legislative Information System. https://lis.virginia.gov/cgi-bin/legp604 .exe?171+ful+CHAP0788. Accessed on September 16, 2017.

Criminal Robots (2017)

As noted in Chapter 2 of this book, one of the most troubling issues about the future of robotics is what the risks are of continued development of artificial intelligence. Will there be a time when robots can "take over the world" or "destroy humankind"? At an intermediary point, the question might be phrased as to whether robots will soon have the ability to carry out somewhat simpler crimes, such as burglary and assault on humans. Christopher Markou, in the Faculty of Law at the University of Cambridge, has thought in some detail about this problem. Here are a few of his most important points as expressed in the blog, The Conversation. Triple asterisks (***) represent the omission of text.

The criminal law has two critical concepts. First, it contains the idea that liability for harm arises whenever harm has been or is likely to be caused by a certain act or omission.

Second, criminal law requires that an accused is culpable for their actions. This is known as a "guilty mind" or "mens rea." The idea behind mens rea is to ensure that the accused both completed the action of assaulting someone and had the intention of harming them, or knew harm was a likely consequence of their action.

So if an advanced autonomous machine commits a crime of its own accord, how should it be treated by the law? How would a lawyer go about demonstrating the "guilty mind" of a non-human? Can this be done be referring to and adapting existing legal principles?

Take driverless cars. Cars drive on roads and there are regulatory frameworks in place to assure that there is a human behind the wheel (at least to some extent). However, once fully autonomous cars arrive there will need to be extensive adjustments to laws and regulations that account for the new types of interactions that will happen between human and machine on the road.

As AI technology evolves, it will eventually reach a state of sophistication that will allow it to bypass human control. As the bypassing of human control becomes more widespread, then the questions about harm, risk, fault and punishment will become more important. Film, television and literature may dwell on the most extreme examples of "robots gone awry" but the legal realities should not be left to Hollywood.

So can robots commit crime? In short: yes. If a robot kills someone, then it has committed a crime (actus reus), but technically only half a crime, as it would be far harder to determine mens rea. How do we know the robot intended to do what it did?

***

Robo-jails?

Even if you solve these legal issues, you are still left with the question of punishment. What's a 30-year jail stretch to an autonomous machine that does not age, grow infirm or miss its loved ones? Unless, of course, it was programmed to "reflect" on its wrongdoing and find a way to rewrite its own code while safely ensconced at Her Majesty's leisure. And what would building "remorse" into machines say about us as their builders?

What we are really talking about when we talk about whether or not robots can commit crimes is "emergence"—where a system does something novel and perhaps good but also unforeseeable, which is why it presents such a problem for law.

***

Source: Markou, Christopher. 2017. "We Could Soon Face a Robot Crimewave . . . the Law Needs to Be Ready." *The Conversation.* https://theconversation.com/we-could-soon-face-a-robot-crimewave-the-law-needs-to-be-ready-75276. Accessed on September 18, 2017. Used by permission of Christopher Markou.

Taylor v. Intuitive Surgical, Inc. (2017)

As robots become a more common part of human life in the 21st century, one might assume that legal cases will arise as to the proper role of such devices in society. Thus far, relatively few court cases of consequence of this nature have arisen. One of the first such cases was heard by the Supreme Court of the state of Washington in 2017. The case involved a reconsideration by the court of lower court decisions about the use of a robotic surgical device by a qualified physician that may have had severe adverse consequences, including the patient's eventual death. The petitioner (the patient's wife) eventually settled with all respondents except for the manufacturer of the robotic device. Her case reached the Supreme Court, which decided that the lower court had not properly considered the manufacturer's legal responsibility in the case. This excerpt describes some of the pertinent issues involved in the case. Triple asterisks (***) indicate the omission of text.

As part of its training, ISI [Intuitive Surgical, Inc.] requires that surgeons perform two proctored surgeries, but hospitals enforce their own requirements for credentialing surgeons to use the da Vinci System. Harrison Medical Center provided credentials after those two proctored procedures. Other hospitals in Washington provided credentials after three or four proctored surgeries. *** ISI recommends that surgeons choose "simple cases" for initial unproctored procedures. *** ISI provided a user's manual to doctors, containing various warnings related to the device. Three warnings are particularly relevant to this case. First, as part of its training, ISI advised surgeons not to perform prostatectomies on obese persons. ISI provided body mass index (BMI) guidelines stating patients should have a BMI of less than 30. Second, ISI advised not to perform prostate procedures on persons who previously underwent lower abdominal surgeries. Third, ISI warned that it was unsafe for the patient not to be in a steep Trendelenburg position (tilted with head downward) during the procedure.

***

After receiving informed consent, Dr. Bildsten performed a robotic prostatectomy on Taylor to treat his prostate cancer using the da Vinci System on September 9, 2008. At the time of surgery, Taylor weighed 280 pounds and had a BMI of 39 (contrary to ISI's advice to choose a patient with a BMI of less than 30). Dr. Bildsten testified that he considered Taylor to be "severely obese." *** Furthermore, Taylor had three prior lower abdominal surgeries (which went against ISI's advice to avoid patients with prior lower abdominal surgeries). During the surgery, Dr. Bildsten did not position Taylor in the steep Trendelenburg position due to his weight (in spite of ISI's advice to conduct the procedure in that position). Although Dr. Bildsten knew that Taylor "was not an optimal candidate," he performed the prostatectomy as his first unproctored procedure using the robotic system. ***

During the surgery, Taylor suffered complications. Dr. Bildsten became aware that Taylor's rectal wall was lacerated. He converted the procedure to an open surgery, and another surgeon came in to fix the rectal tear. Taylor's quality of life was poor after the surgery. He suffered respiratory failure requiring ventilation, renal failure (that ultimately resolved itself), and infection. He was incontinent and had to wear a colostomy bag. He also suffered neuromuscular damage and could no longer walk without assistance. Roughly four years after the surgery, Taylor passed away. A doctor testified that the prostatectomy's complications hastened his death.

***

Conclusion

We find that pursuant to the WPLA [Washington State Product Liability Act], manufacturers have a duty to warn hospitals about the dangers of their products. The manufacturer's warnings to Dr. Bildsten did not excuse its duty to warn Harrison Medical Center. As such, we find that the trial court erred in failing to instruct the jury on this duty. We vacate the jury's defense verdict and remand for a new trial. Further, we hold that the comment k exception in Restatement § 402A is not available to a manufacturer who fails to adequately warn and reverse the trial court's application of negligence in this case.

Source: *Taylor v. Intuitive Surgical Inc.* 2017. Supreme Court of the State of Washington. https://www.courts.wa.gov/opin ions/pdf/922101.pdf. Accessed on September 19, 2017.

Amazon Motion to Quash a Search Warrant (2017)

Although there is relatively little case law dealing with robots, the instances that do exist provide a perspective as to the type of litigation that might arise more commonly in the future. The two examples cited here are of interest because they raise the question as to whether robots have the same First Amendment rights as do humans. In the first case, the court rules that they do. The case is based on a claim by Search King company that Google arbitrarily changed its Page Ranking, a measure of the degree to which page *content matches a search term. Google argued that its robot for performing this ranking had the right to make whatever decision it wanted in this regard because it had the same First Amendment rights as those of a human. In the second instance, Amazon corporation argues that a federal search warrant asking for some of its records is a violation of the First Amendment rights of one of its robots, a cloud-based device known as Alexa Voice Service. Some essential points in the case are as follows. Triple asterisks (***) indicate omitted text.*

From Amazon Motion:

1. Amazon Users' Requests to Alexa Are Protected by the First Amendment

It is well established that the First Amendment protects not only an individual's right to speak, but also his or her "right to receive information and ideas." *** At the heart of that First Amendment protection is the right to browse and purchase expressive materials anonymously, without fear of government discovery. ***

For this reason, courts have recognized that government demands for records of an individual's requests for and purchases of expressive material implicate First Amendment concerns. ***

Here, as the search warrant affidavit recognizes, Amazon customers use their Alexa-enabled devices to request a variety of information and expressive material, including "music playback," "streaming podcasts," and "playing audio books." Given these functions, Amazon's recordings of its customers' requests for information through the devices are subject to the heightened protections of the First Amendment.***

2. Alexa's Responses to User Queries Are Protected by the First Amendment

In addition to the recordings of user requests for information, Alexa's responses are also protected by the First Amendment.

First, as noted above, the responses may contain expressive material, such as a podcast, an audiobook, or music requested by the user. Second, the response itself constitutes Amazon's First Amendment-protected speech. In a similar context, courts have recognized that "the First Amendment protects as speech the results produced by an Internet search engine." *** Alexa's decision about what information to include in its response, like the ranking of search results, is "constitutionally protected opinion" that is "entitled to 'full constitutional protection.'" ***

Absent such protection for customers to seek and obtain expressive content, "the free flow of newsworthy information would be restrained and the public's understanding of important issues and events would be hampered in ways inconsistent with a healthy republic." *** Indeed, the publicity generated by this search warrant in particular has led to numerous articles raising concerns about the use of Alexa-enabled devices and other in-home intelligent personal assistants, and in particular whether use of such devices exposes customers' audio recordings and information requests to government review. *** Such government demands inevitably chill users from exercising their First Amendment rights to seek and receive information and expressive content in the privacy of their own home, conduct which lies at the core of the Constitution. To guard against such a chilling effect, this Court should require the State to make a prima facie showing that it has a compelling need for any recordings that were created as a result of interactions with the Echo device, and that the State's request bears a sufficient nexus to the underlying investigation.

Source: *State of Arkansas v. James A. Bates.* Memorandum of Law in Support of Amazon's Motion to Quash Search Warrant. 2017. Benton County (Arkansas) Circuit Court. https://assets .documentcloud.org/documents/3473799/Alexa.pdf. Accessed on September 19, 2017.

Introduction

Efforts by human inventors to build wondrous machines that perform many of the functions of humans and other animals date back more than two millennia. Over that period of time, the subjects of automata and robots have fascinated writers, artists, engineers, and the common person to an extreme extent. That interest has been expressed in untold numbers of books, articles, electronic articles, and other sources. This chapter can do no more than provide the reader with a hint of the publications available on automata, robots, their applications, problems they pose for humans, and their future in the next century and beyond. The reader should also review the notes for Chapters 1 and 2, which provide additional resources of this type. In some cases, an item may be available in more than one format, such as printed article and Web site. Such instances are so indicated in the listing for those articles.

A group of Pepper customer service robots waits to assist shoppers at the Aeon Mall in Makuhari, Chiba, Japan, in January 2017. The Pepper robot interacts with customers by reading their emotions through facial and vocal expressions, and enhances user experience by learning over time. In addition, the Pepper robot makes fluid movements, has sound recognition, and recharges independently. (Origamiplanemechanic/Dreamstime.com)

Books

Bandyopadhyay, Susmita. 2018. *Intelligent Vehicles and Materials Transportation in the Manufacturing Sector: Emerging Research and Opportunities*. Hershey, PA: IGI Global.

> This book focuses on the use of industrial robots in transportation systems. Chapters deal with topics such as material handling, automated guided vehicles, industrial trucks and cranes, and other types of vehicles and future trends.

Bock, Thomas, and Thomas Linner. 2016. *Construction Robots: Elementary Technologies and Single-Task Construction Robots*. Cambridge, UK: Cambridge University Press.

> This book provides a good general introduction to the types of robots available for construction work, with examples of the many specific devices now in use.

Brockman, John, ed. 2015. *What to Think about Machines That Think: Today's Leading Thinkers on the Age of Machine Intelligence*. New York: Harper Perennial.

> This collection of fascinating articles divulges thoughts about robotics, artificial intelligence, and related topics by a number of the most important thinkers and researchers in the area.

Caldwell, Darwin G., ed. 2013. *Robotics and Automation in the Food Industry: Current and Future Technologies*. Oxford: Woodhead Publishing.

> The author reviews basic principles of robotics and then shows how such principles are used for specific applications in the food industry. Some uses that are discussed include bulk sorting, food chilling and freezing, meat processing, seafood processing, packaging of confectionary products, and applications in the field of fresh food processing.

Calhoun, Laurie. 2016. *We Kill Because We Can: From Soldiering to Assassination in the Drone Age.* London: Zed Books.

> The author provides an overview of the use of robots—primarily drones—in military combat. She explains how such devices are used, possible military consequences, and moral and ethical issues associated with their use.

Callaghan, Vic, James Miller, Roman V. Yampolskiy, and Stuart Armstrong, eds. *The Technological Singularity: Managing the Journey.* Berlin: Springer.

> The papers that make up this anthology cover topics such as "Risks of the Journey to Singularity," "How Changes Agents Can Affect Our Path towards Singularity," "Diminishing Returns and Recursive Self Improving Artificial Intelligence," "The Emotional Nature of Post-Cognitive Singularities," and "Singularity Blog Insights."

Calo, M. Ryan, Michael Froomkin, and Ian Kerr, eds. 2016. *Robot Law.* Cheltenham, UK: Edward Elgar Publishing.

> The seven essays in this anthology deal with the question as to how humans should think about robots as legal entities, the locus of responsibility for robot actions, social and ethical meanings, law enforcement, and war.

Čapek, Karel, David Short, and Arthur Miller. 2011. *Rossum's Universal Robots (R.U.R.): A Collective Drama in Three Acts with a Comedy Prelude.* London: Hesperus.

> This book provides a reprint of Čapek's original play in which the term *robot* was first used.

Caton, Jeffrey L. 2015. *Autonomous Weapon Systems: A Brief Survey of Developmental, Operational, Legal, and Ethical Issues.* Carlisle, PA: Strategic Studies Institute and U.S. Army War College Press. http://publications.armywarcollege.edu/pubs/2378.pdf. Accessed on October 22, 2017.

This excellent publication provides a thorough introduction to the topic of autonomous weapons, with focus on four special aspects of the field: developmental issues, operational issues, legal issues, and ethical issues.

Chopra, Samir, and Laurence F. White. 2011. *A Legal Theory for Autonomous Artificial Agents*. Ann Arbor: University of Michigan Press.

> The five sections in this book deal with legal issues relating to robots and their relationship to basic concepts in law: artificial agents and agency, artificial agents and contracts, attribution of knowledge to artificial agents, tort liability of artificial agents, and personhood for artificial agents.

Ferreira, Maria Isabel Aldinhas, et al., eds. 2017. *A World with Robots: International Conference on Robot Ethics: ICRE 2015*. Cham, Switzerland: Springer International Publishing Imprint.

> This book consists of 17 papers presented at the 2015 International Conference on Robot Ethics. The purpose of the conference was to discuss "the main ethical problems resulting from the widespread adoption of robots." Among the topics discussed were the perceived autonomy of robots, the ethics of robotic search and rescue missions, liability rules for harm caused by robots, safety and ethical concerns associated with the use of driverless cars, and military uses of robots.

Foulkes, Nicholas. 2017. *Automata* (English edition). Paris: Xavier Barral Editeur.

> This book is a richly illustrated, superb history of automata, including descriptions of the way they operated and their use in various aspects of society.

Galliott, Jai. 2015. *Military Robots: Mapping the Moral Landscape*. London: Routledge, Taylor & Francis Group.

> The author begins by reviewing the use of robotic agents in military settings up to the present day. He then reviews

some of the moral issues that are raised by the use of robots in warfare. Finally, he suggests a mechanism by which the "moral responsibility" of a robot can be determined in a military campaign.

Haddadin, Sami. 2014. *Towards Safe Robots: Approaching Asimov's 1st Law.* Heidelberg: Springer.

As robots become more and more a part of human life, work, and research, the issue of safe interactions between them and humans becomes more important. This book discusses ways in which that interaction can be made safer. Individual chapters deal with topics such as crash-testing in robots, competitive robotics, the need for new robotics standards, the role of the robotic coworker, and biomechanics and forensics.

Haikonen, Pentti O. 2012. *Consciousness and Robot Sentience.* Singapore: World Scientific.

As robots become more and more sophisticated, more and more like humans, a fundamental question that becomes more important is how one is to know whether he or she is communicating with a real human or with a robot. This book considers the meaning of *sentience* and *consciousness* and what methods can be used to determine whether or not an entity possesses these qualities, that is, is or is not a human.

Herath, Damith, Christian Kroos, and Stelarc. 2016. *Robots and Art: Exploring an Unlikely Symbiosis.* Singapore: Springer.

The editors point out that this book is "an unusual book" that brings together "science, engineering, and technology on the one side, the arts and critical cultural studies on the other." Some chapters include "We Have Always Been Robots: The History of Robots and Art," "Still and Useless: The Ultimate Animation," "Machines That Make Art," "Android Robots as In-Between Beings," and "Robot Partner—Are Friends Electric."

Hillier, Mary. 1988. *Automata and Mechanical Toys: An Illustrated History*. London: Bloomsbury Book.

> This book provides an excellent illustrated history automata throughout the ages.

Khan, Amar Ali, and Sajid Umair. 2018. *Handbook of Research on Mobile Devices and Smart Gadgets in K-12 Education*. Hershey, PA: Information Science Reference.

> This collection of essays describes some of the rather specific uses of robotics that have been developed for all levels of education from primary through high school.

Krichmar, Jeffrey L., and Hiroaki Wagatsuma, eds. 2011. *Neuromorphic and Brain-Based Robots*. Cambridge, UK: Cambridge University Press.

> This book was written to provide information about the use of robotics for developing a better understanding of the human brain. The essays deal with technical issues, such as the design of robots built on our current understanding of the brain, some technical and philosophical issues arising out of this research, and ethical considerations accompanying this line of research.

Krishna, K. R. 2016. *Push Button Agriculture: Robotics, Drones, Satellite-Guided Soil and Crop Management*. Waretown, NJ: Apple Academic Press.

> This book consists of five (mostly) very long chapters on the aspects of agricultural robotics listed in its title.

Krishna, K. R. 2017. *Agricultural Drones: A Peaceful Pursuit*. Waretown, NJ: Apple Academic Press.

> In contrast to the following citation, this book focuses entirely on the use of drones in agriculture. Some applications include the use of drones to study natural resources and vegetation; deal with soil fertility management, production agronomy, and irrigation and water management

during crop production; to manage weeds in crop fields, crop disease and pest control, and crop yield estimation and forecasting.

Kurzweil, Ray. 2016. *The Singularity Is Near: When Humans Transcend Biology*. London: Duckworth Press.

Kurzweil is one of the most devoted writers about the potential of a technological singularity, in which the mental capacity of robots surpasses that of humans. This book provides a basic introduction to the subject, with an especially interesting chapter on how such an event would affect the human brain, body, and longevity, as well as learning, work, and play.

Lepuschitz, Wilfried, ed. 2018. *Robotics in Education: Latest Results and Developments*. Cham, Switzerland: Springer Publishing.

The papers that make up this book are presentations made at the Eighth International Conference on Robotics in Education held in Sofia, Bulgaria, in April 2017. They include topics such as "Teaching Robotics to Elementary School Children," "Using Robotics to Foster Creativity in Early Gifted Education," "MuseumsBot—An Interdisciplinary Scenario in Robotics Education," "An Elementary Science Class with a Robot Teacher," and "Marine Robotics: An Effective Interdisciplinary Approach to Promote STEM Education."

Lin, Patrick, Keith Abney, and George A. Bekey, eds. 2012. *Robot Ethics: The Ethical and Social Implications of Robotics*. Cambridge, MA: MIT Press.

This book provides a nice general introduction to the moral and ethical issues by developments in robotics. The eight sections deal with introduction and epilogue; design and programming; applications in the military, medicine and care, and psychology and sex; legal issues; and rights and ethics.

Lipson, Hod, and Melba Kurman. 2016. *Driverless: Intelligent Cars and the Road Ahead.* Cambridge, MA: MIT Press.

The authors discuss the evolution of autonomous vehicles, current state of the art, problems that remain to be solved, and some specialized issues associated with the use of driverless cars.

Markowitz, Judith A., ed. 2015. *Robots That Talk and Listen: Technology and Social Impact.* Berlin; Boston: De Gruyter.

The essays in this volume cover a wide range of robot-related topics, such as robots as killers, service robots, comfort robots, the history of automata, robot–human dialogue, applications of robotics in educational programs, and robot feedback.

Marshall, Patrick. 2015. *Robotics and the Economy: Should Workers Fear Increasing Automation?* Washington, DC: CQ Press.

After a good overview of the current status of robots in the workplace, the author considers a series of questions, such as whether or not there is a limit as to what machines can do, the effects of robot workers on the U.S. economy, the role of the government in the development of robotic workers, a history of public concerns about robotic workers, and current research on the role of robots in the workplace.

Miller, Mark R., and Rex Miller. 2017. *Robots and Robotics: Principles, Systems, and Industrial Applications.* New York: McGraw Hill Education.

This book contains an extended introduction to the topic of robotics, in general, along with their many applications. It also provides detailed technical information on various aspects of the design, construction, and operation of industrial robots, such as chapters on sensors and sensing, control methods, and robots and the computer. One chapter is also devoted to an overview of companies currently involved in the design and production of industrial robots.

Murphy, Robin R. 2017. *Disaster Robotics*. Cambridge, MA: MIT Press.

The author writes about the myriad uses of robots in natural and human-made disasters, based on her own experience in more than a dozen such episodes.

Nørskov, Marco, ed. 2016. *Social Robots: Boundaries, Potential, Challenges*. Farnham, Surrey, England; Burlington, VT: Ashgate Publishing.

The essays in this anthology deal with specific aspects of the three general areas mentioned in the title. They include questions of robots and human sexuality, the morality of robots, views of the human social world explicated by robots, and issues of gender in robots.

Pagallo, Ugo. 2013. *The Laws of Robots: Crimes, Contracts, and Torts*. Dordrecht, the Netherlands: Springer.

This book explores most of the traditional fields of law as they apply to robots rather than humans. Included are chapters on crimes, contracts, torts, and robots as legal entities.

Pons, José, ed. 2008. *Wearable Robots: Biomechatronic Exoskeletons*. Chichester, England: Wiley.

This is possibly the best single source available on all aspects of the topic of exoskeletons and wearable robots. It has chapters on a general introduction to the topic; kinematics and dynamics of the devices; human–robot cognitive and physical interaction; technologies available in the field; communication systems for wearable robots; upper, lower, and full-body devices; and a general summary and outlook for the future.

Roberts, Dan. 2014. *Famous Robots & Cyborgs: An Encyclopedia of Robots from TV, Film, Literature, Comics, Toys, and More*. New York: Skyhorse Publishing.

The author provides a very enjoyable and instructive review of the very many ways in which robots have been depicted in the variety of fields indicated in the book's title.

Rosheim, Mark E. 2006. *Leonardo's Lost Robots*. Berlin: Springer. This book provides a detailed and intensive discussion of some of the robots that Leonardo da Vinci designed and may have actually built. The book includes more than 100 technical drawings of these robots.

Russell, Ben, ed. 2017. *Robots: The 500-Year Quest to Make Machines Human*. London: SCALA. The six essays that make up this book discuss topics such as "Being Human: Minds Reflected in Machines"; "Automata, Androids, and Life"; "Pioneers of Cybernetics"; and "Humanoid Robots and the Promise of an Easier Life."

Samani, Hooman, ed. 2016. *Cognitive Robotics*. Boca Raton, FL: CRC Press/Taylor & Francis Group. The essays in this book consider many different aspects of *cognitive robots*, a term used to describe robots with humanlike intelligence. Some of the specific topics included are "When Robots Do Wrong," "Designing Modular AI Robots Inspired by Amerindian Material Culture," "Ethics of Cultural Context and Social Role on Human-Robot Interaction," "A Cognitive Model for Human Willingness to Collaborate with Robots," and "Social Cognition of Robots during Interacting with Humans."

Schweikard, Achim, and Floris Ernst. 2015. *Medical Robotics*. Berlin: Springer International. This book provides a comprehensive review of major topics in the field of medical robots, with about a dozen chapters on topics such as spatial position and orientation, robot kinematics, joint velocities, navigation and registration, treatment planning, motion replication, and applications of surgical robots.

Shanahan, Murray. 2015. *The Technological Singularity.* Cambridge, MA: MIT Press.

The author explains the concept of technological singularity and suggests a number of scenarios that might result if such an event were actually to occur.

Siciliano, Bruno, and Oussama Khatib, eds. 2016. *Springer Handbook of Robotics,* 2nd ed. Berlin: Springer.

This book qualifies as one of the most comprehensive introductions to the field of robotics currently available. Essays by more than 100 experts in the field deal with a range of topics divided into seven main sections, each consisting of more than a dozen essays: the foundations of robotics, robot design, sensing and perception, manipulation and interfaces, moving in the environment, robots at work, and robots and humans.

Tchoń, Krzysztof, and Wojciech Gasparski, eds. 2014. *A Treatise on Good Robots.* New Brunswick, NJ: Transaction Publishers.

The essays that make up this volume discuss some of the positive results that can be expected from expanded use of robots in the future, such as care of the elderly, maintenance activities, and new systems of navigation.

Thompson, Steven John, ed. 2018. *Androids, Cyborgs, and Robots in Contemporary Culture and Society.* Hershey, PA: IGI Global, Engineering Science Reference.

This selection of essays deals with a rather broad range of robot-related topics, such as legal issues relating to robotic use, the role of androids and cyborgs in Japanese comics and animations, pros and cons of the construction of cyborgs, and the question as to whether robots can feel pain or not.

Troccaz, Joselyne. 2012. *Medical Robotics.* Hoboken, NJ: Wiley ISTE.

The 10 sections of this book contain chapters on topics such as characteristics and state of the art, medical robotics in the service of the patient, augmented reality, design of medical robots, vision-based control, and tele-manipulation.

Truitt, Elly Rachel. 2015. *Medieval Robots: Mechanism, Magic, Nature, and Art*. Philadelphia: University of Pennsylvania Press.
This superb book discusses not only the technical aspects of automata but also their religious, social, educational, and other purposes.

Van Wynsberghe, Aimee. 2015. *Healthcare Robots: Ethics, Design and Implementation*. Farnham, Surrey; Burlington, VT: Ashgate.
This book provides a useful introduction to the ways in which robots can be used in health care systems, some potential problems associated with such use, and the future of robots in health care.

Wadhwa, Vivek. 2017. *Driver in the Driverless Car*. Oakland, CA: Berrett-Koehler Publishers.
The author considers four general areas of concern about driverless cars: the way they have and are changing the world in which we live, the question as to whether this new technology will benefit everyone equally, the risks and rewards of the new technology, and the issue of autonomy versus dependency in the rise of driverless cars.

Wallach, Wendell, and Peter Mario Asaro, eds. 2016. *Machine Ethics and Robot Ethics*. Aldershot, Hampshire: Ashgate Publishing.
This publication consists of a collection of articles that have previously appeared in other peer-reviewed journals. They fall into five major categories: foundations of robotics and artificial intelligence, robotic ethics, machine ethics, moral agents and agency, and law and policy.

Weaver, John Frank. 2014. *Robots Are People Too: How Siri, Google Car, and Artificial Intelligence Will Force Us to Change Our Laws*. Santa Barbara, CA: Praeger.

The nine chapters in this book deal with modern interpretations of Asimov's Three Laws for Robots and ask whether a robot may hurt a human being, whether a robot must always obey orders from a human, and whether a robot can or should protect itself.

Wosk, Julie. 2015. *My Fair Ladies: Female Robots, Androids, and Other Artificial Eves*. New Brunswick, NJ: Rutgers University Press. The author reviews the history of female robots, androids, automata, cyborgs, and other artificial females. In addition to descriptions of the machines, she discusses the significance of this line of research for societal views about women in general.

Xie, Shane. 2016. *Advanced Robotics for Medical Rehabilitation: Current State of the Art and Recent Advances*. Cham: Springer International Publishing. This somewhat technical publication reviews recent developments and future prospects for the use of robotics in medical rehabilitation programs.

Zhang, Dan, and Bie Wei, eds. 2018. *Robotics and Mechatronics for Agriculture*. Boca Raton, FL: CRC Press; Taylor & Francis Group. Some of the applications of robotics in the field of agriculture discussed in this book are composting systems, weed management, unstructured agricultural environments, smart cameras in agriculture, and cooperative robotic systems.

Articles

Some journals are devoted entirely to robot-related topics. They vary in area of interest and technical level. They include the following journals:

Advanced Robotics: ISSN: 0169-1864; eISSN: 1568-5535.
Autonomous Robots: ISSN: 0929-5593; eISSN: 1573-7527.

International Journal of Advanced Manufacturing Technology: ISSN: 0268-3768; eISSN: 1433-3015.

International Journal of Humanoid Robotics: ISSN: 0219-8436; eISSN: 1793-6942.

International Journal of Robotics and Mechatronics: ISSN: 2288-5889.

International Journal of Robotics Research: ISSN: 02783649; eISSN: 17413176.

Journal of Field Robotics: eISSN: 1556-4967.

Journal of Intelligent and Robotic Systems: ISSN: 0921-0296; eISSN: 1573-0409.

Journal of Robotic Surgery: eISSN: 1863-2491.

Journal of Robotics and Mechatronics: ISSN: 0915-3942; eISSN: 1883-8049.

Paladyn, Journal of Behavioral Robotics: ISSN: 2080-9778; eISSN: 2081-4836.

Robotics: ISSN: 2218-6581.

Robotics and Autonomous Systems: ISSN: 0921-8890.

Robots and Computer-Integrated Manufacturing: ISSN: 0736-5845.

Alesich, Simone, and Michael Rigby. 2017. "Gendered Robots: Implications for Our Humanoid Future." *IEEE Technology and Society Magazine* 36(2): 50–59.

> The authors consider how gender is a factor in the development and operation of robots, both for the robots produced and for their human creators. They suggest that gendered robots are likely to provide humans with a new and different understanding of the nature of gender among humans.

Andersen, Espen, et al. 2014. "The Technological Singularity." *Ubiquity* 2014: 1–8.

> This article is the opening statement for a series of articles in this and later issues of the journal on the technological singularity. A complete list of all articles in the

symposium and links to those articles can be found at Andersen's web page, "Applied Abstractions," https://appliedabstractions.com/2014/10/20/acm-ubiquitys-singularity-symposium/.

Anderson, Susan Leigh. 2008. "Asimov's 'Three Laws of Robotics' and Machine Metaethics." *AI and Society* 22(4): 477–493.

The author uses Isaac Asimov's novelette on robotics to consider issues surrounding current efforts to develop an "ethical" robot. She reviews some of the problems involved in such research and ways in which those problems might be dealt with.

Arbib, Michael A., and Jean-Marc Fellous. 2004. "Emotions: From Brain to Robot." *Trends in Cognitive Sciences* 8(12): 554–561.

The authors review attempts to provide robots with human-like emotions for the purpose of carrying on conversations with humans. They inquire into similar possible efforts in which a robot's "emotions" are used to determine the machines' behaviors.

Bacon, Simon. 2013. "We Can Rebuild Him!: The Essentialisation of the Human/Cyborg Interface in the Twenty-First Century, or Whatever Happened to the Six Million Dollar Man?" *AI & Society* 28(3): 267–276.

This article compares public perceptions of a robotic future as displayed in the late 20th century (optimistic and promising) to that of the early 20th century (wary and pessimistic).

Ball, David, et al., eds. 2017. "JFR Special Issue on Agricultural Robotics." *Journal of Field Robotics* 34(6): 1037–1038.

This special edition of the journal focuses on many aspects of agricultural robots, with articles on specific applications such as fruit detection and yield estimation in fruit orchards, rice harvesting, basic studies on row crops, pruning of grape vines, and harvesting sweet peppers.

Bdiwi Mohamad, Marko Pfeifer, and Andreas Sterzing. 2017. "A New Strategy for Ensuring Human Safety during Various Levels of Interaction with Industrial Robots." *CIRP Annals— Manufacturing Technology* 66(1): 453–456. http://www.science direct.com/science/article/pii/S0007850617300094. Accessed on October 24, 2017.

> This paper addresses an ongoing issue in the use of industrial robots: the possible risks to humans in their interactions with robots. The authors review the research on this issue and suggest a new approach to reducing the risks to humans in such interactions.

Billard, Aude. 2017. "On the Mechanical, Cognitive and Sociable Facets of Human Compliance and Their Robotic Counterparts." *Robotics and Autonomous Systems* 88: 157–164.

> Billard notes that most research thus far has focused on the first two of these three characteristics (mechanical and cognitive interactions) but that the third factor (social) may be at least as important as these two. She suggests that it may be possible that humans can learn from robots, rather than the other way around, in this regard.

Bogue, Robert. 2017. "Europe Leads the Way in Assistive Robots for the Elderly." *Industrial Robot* 44(3): 253–258.

> The author takes note of the increasing need for elderly care in Europe and writes of developing efforts to use robotics to deal with the issue. He outlines some of the programs currently in effect and those under consideration.

Bogue, Robert. 2017. "Sensors Key to Advances in Precision Agriculture." *Sensor Review* 37(1): 1–6.

> The author points out the importance of sensors in agricultural applications of robotics and outlines some of the requirements of such systems.

Brain-Inspired Intelligent Robotics: The Intersection of Robotics and Neuroscience. 2016. Special supplement to the journal *Science.*

http://www.sciencemag.org/sites/default/files/custom-publish ing/documents/Brain-inspired-robotics-supplement_final.pdf. Accessed on October 22, 2017.

This excellent overview of the relationship of neurosciences to robotics is an introduction to the journal's new speciality journal *Science Robotics*. The supplement contains articles such as "Creating More Intelligent Robots through Brain-Inspired Computing"; "Deep Learning: Mathematics and Neuroscience"; "Collective Robots: Architecture, Cognitive Behavior, Model, and Robot Operating System"; "Toward Robust Visual Cognition through Brain-Inspired Computing"; "Brain-Like Control and Adaptation for Intelligent Robots"; and "Neuro-robotics: A Strategic Pillar of the Human Brain Project."

Chang, Chih-Wei, et al. 2010. "Exploring the Possibility of Using Humanoid Robots as Instructional Tools for Teaching a Second Language in Primary School." *Educational Technology & Society* 13(2): 13–24.

Many journal articles are available on the topic of educational robotics. This listing provides an example of the types of questions and situations discussed in such papers.

Chen, Steven W., et al. 2017. "Counting Apples and Oranges with Deep Learning: A Data-Driven Approach." *IEEE Robotics and Automation Letters* 2(2): 781–788.

This article provides an interesting explanation of one very specific application of robotics in the field of agriculture.

Collingwood, Lisa. 2017. "Privacy Implications and Liability Issues of Autonomous Vehicles." *Information & Communications Technology Law* 26(1): 32–45.

The author notes that much research on autonomous vehicles has focused on technical problems: will they work and what problems are they likely to cause for drivers and the community as a whole. In this article, she focuses on privacy implications and legal issues relating to driverless vehicles.

Danaher, John. 2016. "Robots, Law and the Retribution Gap." *Ethics and Information Technology* 18(4): 299–309.
This article discusses one of the many legal issues that are arising as robots become a more pervasive part of human life, retribution. The question is whom one can blame for mishaps when the mishap results from robotic behavior.

Dautenhahn, Kerstin, et al. 2009. "Robots in the Wild: Exploring Human-Robot Interaction in Naturalistic Environments." *Interaction Studies* 10(3): 269–273.
This issue of the journal is devoted entirely to a discussion of the possible interactions between humans and robots in a variety of settings, such as search and rescue operations, museums and other public spaces, therapeutic and educational settings, domestic activities, artistic contexts, and medical and health care facilities.

Dreier, Thomas, and Indra Spiecker genannt Döhmann. 2012. "Legal Aspects of Service Robotics." *Poiesis & Praxis* 9(3–4): 201–217.
The authors offer a good introduction to the way in which legal systems deal with technological problems before discussing the more specific issues of the types of legal issues currently associated with the development and use of service robots.

Dyrkolbotn, Sjur. 2017. "A Typology of Liability Rules for Robot Harms." *Intelligent Systems, Control and Automation: Science and Engineering* 84: 119–133.
The author suggests that the development of robots with artificial intelligence has caused legal experts to view traditional legal theories in new and different ways. He attempts to provide a framework for that new approach to robotics law.

Evan, Dave. 2016. "Is Mankind on the Way towards Technological Singularity?" *Security Index* 20(2): 23–29.

> This article is the report of an interview with the listed author, chief futurist and senior director at Cisco Internet Business Solutions Group, on his views on the possible future of a technological singularity.

Faragó, Tamás, et al. 2014. "Social Behaviours in Dog-Owner Interactions Can Serve as a Model for Designing Social Robots." *Interaction Studies* 15(2): 143–172.

> The authors explain how the normal behavior of dogs as pets can be used as a model for the development of social behaviors in robots.

Galliott, Jai, and Tim McFarland. 2016. "Autonomous Systems in a Military Context: A Survey of the Legal Issues." *International Journal of Robotics Applications and Technologies* 4(2): 34–52 and 53–68.

> This pair of papers discuss a range of ethical issues associated with the military use of robots, such as legal obligations of military forces, the ability of those forces to meet such obligations, relevant domestic laws, special issues related to the use of drones, international laws and policies, arms control laws and treaties, international human rights laws, what conditions justify the onset of armed conflict, who the legitimate targets may be in such a conflict, and methods for dealing with violations of international and national laws for armed conflict.

Gerdes, Anne. 2014. "Ethical Issues Concerning Lethal Autonomous Robots in Warfare." *Frontiers in Artificial Intelligence and Applications* 273: 277–289.

> With the increasing use of robots in military engagements has come an increasing concern about ethical and moral issues related to such applications. The author begins her essay with a review of the current use of robots in warfare.

She then proposes a way of testing, a Moral Military Turning Test, by which the moral actions of robots can be tested.

Ghezzi, Tiago Leal, and Oly Campos Corleta. 2016. "30 Years of Robotic Surgery." *World Journal of Surgery* 40(10): 2550–2557. https://link.springer.com/article/10.1007%2Fs00268-016-3543-9. Accessed on October 21, 2017.
The authors review the development of robotic surgery, the types of such surgery currently available, some problems associated with robotic surgery, and examples of the devices available in the field.

Hori, Koichi. 2015. "How Should We Advance the Research and Development of AI, Watching the Technological Singularity?" *Journal of Information Processing and Management* 58(4): 250–258.
The author takes note of the theory of technological singularity and asks how such a potential future for humankind can and should affect current and future research on artificial intelligence.

"Industrial Robot Integration." 2017. RobotWorx. https://www.robots.com/applications. Accessed on October 24, 2017.
This Web site is a superb introduction to the many types of work that can be done by industrial robots. It is divided into three major sections: welding robots, material handling robots, and other types of industrial robots. Each of these sections then provides links to more than a dozen specific applications, such as (for welding) arc welding, orbital welding, plasma cutting, resistance welding, electron beam, laser welding, and oxyacetylene welding. Video presentations of the technique are provided for each specific type of job listed.

Jahanshahi, Mohammad R., et al. 2017. "Reconfigurable Swarm Robots for Structural Health Monitoring: A Brief Review."

International Journal of Intelligent Robotics and Applications 1(3): 287–305.

Continuous inspection of infrastructure, such as bridges and buildings, is essential for their ongoing safety and possible maintenance needs. But accessing many such structures can be a physically difficult task for humans and other types of machines. This article explains how swarm robots can conduct the necessary inspections at much lower cost and greater safety than traditional systems. (For more information on swarm robots in general, see Navarro and Matía 2013.)

Johansson, Linda. 2010. "The Functional Morality of Robots." *International Journal of Technoethics* 1(4): 65–73.

The author argues that robots should be evaluated for their moral status in the same way that humans are judged to be moral or not. She suggests a version of the original Turing test (called the Moral Turing Test) can be used to make such judgments.

Kahn, Peter H., Jr., Healther E. Gary, and Solace Shen. 2013. "Children's Social Relationships with Current and Near-Future Robots." *Child Development Perspectives* 7(1): 32–37.

The authors note that young children today are growing up in educational settings in which robots are more likely to present in a teaching or assisting role. They ask how these students will view robotic involvement and how those views will affect their learning experiences.

Koopman, Philip, and Michael Wagner. 2017. "Autonomous Vehicle Safety: An Interdisciplinary Challenge." *IEEE Intelligent Transportation Systems Magazine* 9(1): 90–96. Preprint version available at https://users.ece.cmu.edu/~koopman/pubs/koopman 17_ITS_av_safety.pdf. Accessed on October 21, 2017.

The authors consider the range of issues associated with the use of driverless vehicles, including efficiency of technology, training of human drivers, legal responsibilities, and regulatory issues.

Leung, Tiffany, and Dinesh Vyas. 2014. "Robotic Surgery: Applications." *American Journal of Robotic Surgery* 19(1): 38–41. https://www.ncbi.nlm.nih.gov/pmc/articles/PMC4615607/. Accessed on October 21, 2017.
  This article provides a general overview of the evolution of robotic surgery, some fields in which it is used, and the economics and ethics of the technology.

Miller, Lantz Fleming. 2015. "Granting Automata Human Rights: Challenge to a Basis of Full-Rights." *Human Rights Review* 16(4): 369–391.
  The author notes that discussions about the possibility of "robot rights" have not yet begun in earnest. He attempts to guide those discussions on the basis of two fundamental questions: Do humans have the obligation to grant those rights to robots, and how would a system of robot rights compare to comparable systems of human rights?

Navarro, Iñaki, and Fernando Matía. 2013. "An Introduction to Swarm Robotics." *ISRN Robotics* 2013(1): 1–10. https://www.hindawi.com/journals/isrn/2013/608164/. Accessed on October 25, 2017.
  Swarm robots are collections of small identical devices whose behaviors are homogeneous and controllable by outside agents. This article provides basic information about their technical basis and some potential uses. (Also see Jahanshahi et al. 2017 for one such application.)

Oberson, X. 2017. "Taxing Robots? From the Emerge of an Electronic Ability to Pay to a Tax on Robots or the Use of Robots." *World Tax Journal* 9(2): 247–261.
  The author notes that, as robots become more human like and take on many of the actions of humans, there are circumstances in which such machines might be subject

to taxes. He outlines some of those situations and ways in which taxing could be provided.

Rossi, Cesare, and Flavio Russo. 2017. "Automata (towards Automation and Robots)." *History of Mechanism and Machine Science* 33: 353–380. https://link.springer.com/chapter/10.1007/978-3-319-44476-5_21. Accessed on October 21, 2017.
This article provides a general overview of the history of automata and robots from ancient time to the 20th century.

Shah, Jay, Arpita Vyas, and Dinesh Vyas. 2014. "The History of Robotics in Surgical Specialties." *American Journal of Robotic Surgery*. 1(1): 12–20.
The major focus of this article is a discussion of the types of procedures in which robotic surgery is now performed, including the fields of otolaryngological, neurosurgery, gynecology, cardiothoracic surgery, gastroenterology, urology, and orthopedics. Some discussion of the future of robotic surgery is included.

Sharkey, Amanda J. C. 2016. "Should We Welcome Robot Teachers." *Ethics and Information Technology* 18(4): 283–297. https://link.springer.com/article/10.1007/s10676-016-9387-z. Accessed on October 23, 2017.
The author begins by identifying the four major roles that robots may play in an educational setting: classroom teacher, companion and peer, care companion, and telepresence teacher. She then discusses ethical issues that may arise within each of these settings that involve questions of privacy, attachment, deception, and loss of human contact.

Shaw, Jonathan. 2017. "The RoboBee Collective." *Harvard Magazine*. https://harvardmagazine.com/2017/11/robobee-harvard. Accessed on October 24, 2017.
This article describes the evolution of research on insect-like robots that can be used for research on honeybees,

as well as potential applications such as crop pollination, studies of weather patterns, search and rescue operations, and weather and climate research.

Silva, Manuel, and J. A. Tenreiro Machado. 2007. "A Historical Perspective of Legged Robots." *Journal of Vibration and Control* 13(9–10): 1447–1486.
    The authors provide a review of the evolution of legged robots, a discussion of the present state of the art for such devices, and some analysis of problems still to be solved in the further development of legged robots.

Sparrow, Robert. 2016. "Robots in Aged Care: A Dystopian Future?" *AI & Society* 31(4): 445–454.
    The author envisions a future in which care of the elderly becomes entirely the responsibility of robots, with the negative result stated in the title of the article. He outlines some of the reasons he believes such a plan would result in more negative outcomes than positive results.

Spennemann, Dirk H. R. 2007. "On the Cultural Heritage of Robots." *International Journal of Heritage Studies* 13(1): 4–21.
    All societies possess a collection of beliefs, physical artifacts, and other remnants by which future civilizations can understand their culture. Such remnants are often lost but often retained and honored by future societies. The author asks here whether or not the cultural heritage of robots will be among those that will be remembered in the future, and, if such a pathway is desired, what actions should be taken now to ensure that the robotic cultural record is retained in an accurate form.

Wareham, Christopher. 2011. "On the Moral Equality of Artificial Agents." *International Journal of Technoethics* 2(1): 35–42.
    The author points out that as the sophistication of robots grow, so does their involvement in moral and ethical issues. He suggests a mechanism for judging the moral status of an "artificial agent" (robot).

Warwick, Kevin, and Human Shah. 2014. "The Turing Test: A New Appraisal." *International Journal of Synthetic Emotions* 5(1): 31–45.

The authors discuss the Turing test, as it was originally conceived by British mathematician Alan Turing in 1950. They then consider ways in which that test can be used in its original or modified form to distinguish between human and robot in a conversation between one pair of the same or opposite entities.

Ya-Hue, Wu, Christine Fassert, and Anne-Sophie Rigaud. 2012. "Designing Robots for the Elderly: Appearance Issue and Beyond." *Archives of Gerontology and Geriatrics* 54(1): 121–126.

This paper reports on a study of reactions by elderly individuals with mild cognitive impairment to the use of robots as home companions. (They were generally not enthusiastic about the robots.) Participants in the study were then asked to discuss their reactions, in general, to the very idea of using robots for such a purpose.

Zielinska, Teresa. 2016. "Professional and Personal Service Robots." *International Journal of Robotics Applications and Technologies* 4(1): 63–82.

The author provides a short history of robotics and explains the difference among terms such as *service robot, personal service robot,* and *professional service robot,* along with a description of the activities carried out by each.

Reports

Note that some of these reports are available only by purchase. The listings here provide sufficient information to understand the major findings of such reports, as well as links through which the report can be accessed.

"Agricultural Robots Market Analysis by Product (UAV, Driverless Tractor, Milking Robots, Materials Management), by Application (Field Farming, Dairy, Animal, Soil & Crop Management), & Segment Forecasts, 2014–2025." 2017. Grand View Research. Summary and ordering information at http://www.grandviewresearch.com/industry-analysis/agricultural-robots-market. Accessed on October 23, 2017.

This report is an essential resource for anyone interested in the current status and projected future of the use of robots in agriculture.

Christian, Russell. 2016. "Making the Case: The Dangers of Killer Robots and the Need for a Preemptive Ban." Human Rights Watch. https://www.hrw.org/report/2016/12/09/making-case/dangers-killer-robots-and-need-preemptive-ban. Accessed on September 24, 2017.

This report updates two previous HRW reports on the use of robots as killing devices, the threats they pose, and some steps that can be taken to deal with these threats: an excellent general summary and discussion of the issue.

"Digital Life in 2025: AI, Robotics, and the Future of Jobs." 2014. Pew Research Center. http://www.pewinternet.org/2014/08/06/future-of-jobs/. Accessed on July 17, 2017.

This report summarizes the responses of 1,896 experts in the field of robotics to the question: "The economic impact of robotic advances and AI—Self-driving cars, intelligent digital agents that can act for you, and robots are advancing rapidly. Will networked, automated, artificial intelligence (AI) applications and robotic devices have displaced more jobs than they have created by 2025?" Responses were divided almost equally between those who thought robots would displace large numbers of human workers and those who believed that the development of robots would have essentially no net effect on the world's workforce.

Domsalla, Matthew R. 2012. "Rise of the Ethical Machines." Maxwell Air Force Base, AL: School of Advanced Air and

Space Studies. http://www.dtic.mil/docs/citations/AD1019463. Accessed on October 22, 2017.

> This "examines the moral, ethical, and legal issues surrounding the development of autonomous systems capable of employing lethal force." It begins by discussing international law related to the use of such devices and then considers the problems involved in developing such systems that are capable of ethical reasoning.

"Exoskeleton Report." 2017. http://exoskeletonreport.com/. Accessed on October 22, 2017.

> This Web site is devoted entirely to the topic of exoskeleton. It provides information on current technology, types of devices available, status of the industry, books on exoskeletons and wearable robotics, presentations on the topic, and upcoming events.

"From Internet to Robotics. 2016 Edition." 2016. University of California San Diego, et al. http://jacobsschool.ucsd.edu/contextualrobotics/docs/rm3-final-rs.pdf. Accessed on October 25, 2017.

> This report is the latest update of a series of reports about the role of robotics in American society produced by teams of experts from industry and academia. The report covers a wide range of topics, including robots in manufacturing and the supply chain; consumer and professional services; health, independence, and quality of life; public safety; earth sciences and space research; workforce development; and legal, ethical, and economic issues.

Ghaffarzadeh, Khasha. 2017. "Agricultural Robots and Drones 2017–2027: Technologies, Markets, Players." IDTechEx. https://www.idtechex.com/research/reports/agricultural-robots-and-drones-2017-2027-technologies-markets-players-000525.asp. Accessed on October 20, 2017.

> As the title suggests, this report provides an overview and some predictions about the roles that robots can be

expected to be playing in the field of agriculture over the next decade.

Gownder, J. P., et al. "The Future of Jobs, 2027: Working Side by Side with Robots." Forrester. https://www.forrester.com/report/ The+Future+Of+Jobs+2027+Working+Side+By+Side+With+ Robots/-/E-RES119861. Accessed on October 20, 2017.

This report considers fears about robots "stealing" jobs from humans, the anticipated division of labor between humans and robots, the likelihood that robots will transform the nature of work rather than simply replacing humans in jobs, and the probability that robots in the workplace will transform social institutions as we know them.

"Medical Robots Market by Product (Instruments & Accessories and Robot Systems (Surgical Robots, Rehabilitation Robots, Hospital Robots, Non-Invasive Surgery Robots), Application (Orthopedic, Laparoscopy, Neurology)—Global Forecasts to 2021)." 2017. http://www.marketsandmarkets.com/Market-Reports/medical-robotic-systems-market-2916860.html. Accessed on October 20, 2017.

This very detailed report outlines the business future for medical robots of many types over the next five years.

"Preliminary Draft Report of COMEST on Robotics Ethics." 2016. UNESCO. World Commission on the Ethics of Scientific Knowledge and Technology (COMEST), http://unesdoc.unesco .org/images/0024/002455/245532E.pdf. Accessed on October 22, 2017.

This document reviews the applications of robots in a wide variety of fields and asks what rights they should have, if any, in each of these fields. The commission makes no recommendations in this report, although such a section may be developed for the final report.

Sharkely, Noel, et al. 2017. "Our Sexual Future with Robots." Foundation for Responsible Robotics. https://responsiblerobotics

.org/2017/07/05/frr-report-our-sexual-future-with-robots/.
Accessed on March, 17, 2018.

The Foundation for Responsible Robotics studied the
question of the role of robots in human sexuality in the
future. After a review of the current state of research in that
field, researchers asked seven questions about the topic:

1. Would people have sex with a robot?
2. What kind of relationship can we have with a robot?
3. Will robot sex workers and bordellos be acceptable?
4. Will sex robots change societal perceptions of gender?
5. Could sexual intimacy with robots lead to greater
   social isolation?
6. Could robots help with sexual healing and therapy?
7. Would sex robots help to reduce sex crimes?

They then discuss the implications of the responses
received to these questions.

Smith, Aaron, and Monica Anderson. 2017. "Automation in
Everyday Life." Pew Research Center. http://www.pewinternet
.org/2017/10/04/americans-attitudes-toward-a-future-in-which-
robots-and-computers-can-do-many-human-jobs/. Accessed on
October 20, 2017.

This Pew Research poll of American attitudes about the
role of robots in the workplace in the future found that
about a quarter of respondents thought it highly likely
that robots would be taking over human jobs in the future
and that 73 percent of respondents were either "very" or
"somewhat" worried about this type of change in the
workplace.

The poll also asked about attitudes toward the use of
robots as caregivers in the future. Sixty-five percent of
respondents had not even heard about such an option,

but more than half thought that such an application of robots was "very" or "somewhat" realistic. About 40 percent of respondents said that they would be interested in having a robot as a caregiver, although this fraction varied considerably with age, education, and income.

Stubbings, Carol, and Jon Williams. 2017. PwC. http://www .pwc.com/us/en/hr-management/publications/assets/pwc-workforce-of-the-future-the-competing-forces-shaping-2030 .pdf. Accessed on October 20, 2017.

> This report attempts to predict the world of work as it will exist in about 2030. It envisions four possible scenarios, a "yellow" world, in which "humans come first"; a "red" world, in which "innovation rules"; a "green" world in which "companies care"; and a "blue" world, in which "corporate is king." The authors then discuss in detail the anticipated characteristics of each of these scenarios.

"Workshop on the Future of Social Protection." 2017. http:// www.oecd.org/els/soc/Agenda-workshop-on-the-future-of-social-protection-Berlin-12-June-2017.pdf. Accessed on October 18, 2017.

> This conference was held in Berlin on June 12, 2017, to consider various ways of dealing with changes in work patterns as the result of developing in the future. The web page lists the addresses given at the conference, with a link to each of those speeches.

"World Robotics Report 2016." 2016. International Federation of Robotics. https://ifr.org/ifr-press-releases/news/world-robot ics-report-2016. Accessed on October 20, 2017.

> This annual report by the IFR summarizes currently available data and statistics on the production and use of robots throughout the world. This Web site provides the press release for the report and contains links to the major parts of the report.

Internet

"Advantages and Disadvantages of Automating with Industrial Robots." 2017. Robotworx. https://www.robots.com/blog/view ing/advantages-and-disadvantages-of-automating-with-indus trial-robots. Accessed on August 20, 2017.

> This article points out that the use of robots can provide a number of benefits for an industry but that such use also involves certain risks, both of which are mentioned and discussed in the article.

Alam, Nafis. 2017. "Will Robots Rule Finance?" Discover. http:// blogs.discovermagazine.com/crux/2017/06/29/will-robots-rule- finance/#.WckcC7KGOpp. Accessed on September 25, 2017.

> This article reviews the type of tasks that robots can now and will be able in the future to perform in the field of finance. It also provides some discussion of the variety of problems that may arise as a consequence of robot- directed decisions in various fields of finance.

Al-Hasani, Salim. 2017. "800 Years Later: In Memory of Al-Jazari, a Genius Mechanical Engineer." Muslim Heritage. http:// muslimheritage.com/article/800-years-later-memory-al-jazari- genius-mechanical-engineer. Accessed on October 27, 2017.

> This article provides an excellent introduction to and over- view of the life and career of perhaps the most famous of all Islamic technicians and engineers. It includes descrip- tions of some of the automata invented by al-Jazari.

Andersen, Kurt. 2014. "Enthusiasts and Skeptics Debate Artificial Intelligence." *Vanity Fair*. https://www.vanityfair.com/news/ tech/2014/11/artificial-intelligence-singularity-theory. Accessed on October 18, 2017.

> The author reports on interviews with specialists who are concerned about a possible technological singularity in the not-too-distant future and those who are convinced such an event will never occur.

Anderson, Kenneth, and Matthew Waxman. 2013. "Law and Ethics for Autonomous Weapon Systems. Why a Ban Won't Work and How the Laws of War Can." Hoover Institution. Stanford University. http://media.hoover.org/sites/default/files/documents/Anderson-Waxman_LawAndEthics_r2_FINAL.pdf. Accessed on October 17, 2017.

> This long and carefully reasoned article suggests that the world's nations need to rely on traditional and existing legal methods for controlling the use of robots in military situations rather than banning the machines outright.

"Autonomous Flying Microrobots (RoboBees)." 2017. Wyss Institute. https://wyss.harvard.edu/technology/autonomous-flying-microrobots-robobees/. Accessed on October 24, 2017.

> This web page describes a new line of research on robots built on the size of insects that operate on the same general principles as insects. They seem to have promise in the areas of crop pollination, search and rescue missions, surveillance, high-resolution weather studies, and climate and environmental monitoring.

Bascetta, Luca, Marco Baur, and Giambattista Gruosso. 2017. "'ROBI': A Prototype Mobile Manipulator for Agricultural Applications." *Electronics* 6(2): 39. doi:10.3390/electronics6020039. http://www.mdpi.com/2079-9292/6/2/39. Accessed on March 17, 2018.

> This article provides an interesting and understandable explanation of the development of a new type of robot for use in a variety of agricultural applications.

Bernard, Tara Siegel. 2016. "The Pros and Cons of Using a Robot as an Investment Adviser." *New York Times*. https://www.nytimes.com/2016/04/30/your-money/the-pros-and-cons-of-using-a-robot-as-an-investment-adviser.html?mcubz=3&_r=1. Accessed on October 18, 2017.

> This article describes the use of so-called robo-advisors in making investments and discusses some of the advantages

and disadvantages and risks and benefits on relying on this type of investing.

Bertolini, Andrea. 2017. "The Legal Issues of Robotics." *Robohub*. http://robohub.org/the-legal-issues-of-robotics/. Accessed on October 22, 2017.
>   The author notes that the major legal issues involving the use of robots concern liability in case of accident or damage and privacy issues of those working with robots. This essay includes some of the author's responses to adoption of the 2017 resolution on robotics by the European Parliament.

Calo, Ryan. 2016. "When a Robot Kills, Is It Murder or Product Liability?" http://www.slate.com/articles/technology/future_tense/2016/04/a_robotics_law_expert_on_paolo_bacigalupi_s_mika_model.html. Accessed on October 22, 2017.
>   A Stanford University law professor reflects on a recent fiction book in which a robot kills a human. He asks who has the legal and moral responsibility for the act.

Captain, Shawn. 2016. "Robots Are Developing Feelings. Will They Ever Become 'People'?" Fast Company. https://www.fastcompany.com/3062868/robots-are-developing-feelings-will-they-ever-become-people. Accessed on October 22, 2017.
>   Captain begins by reviewing past science fiction in which robots have exhibited feelings or emotions. He then asks whether such a development is ever really possible and, if so, how a robot's feelings and emotions will differ from those of a human, if at all.

"Category Archives: Driverless Cars." 2017. Dashboard Insights. Foley & Lardner LLP. https://www.autoindustrylawblog.com/driverless-cars/. Accessed on October 21, 2017.
>   This Web site lists a number of blog articles on all aspects of the development and use of autonomous vehicles.

Choi, Charles Q. 2008. "Humans Marrying Robots? A Q&A with David Levy." *Scientific American.* https://www.scientificamerican.com/article/humans-marrying-robots/. Accessed on October 20, 2017.

> The author interviews a pioneer in research on artificial intelligence, who sees no reason to believe that marriages between humans and robots will not be legal by the middle of the 21st century.

Cowen, Tyler. 2017. "Let Robots Teach American Schoolkids." *Bloomberg View.* https://www.bloomberg.com/view/articles/2017-07-17/let-robots-teach-american-schoolkids. Accessed on October 23, 2017.

> The author argues that "the nation's whole K-12 education system is artificial, so why not give automatons a chance?" He reviews the way in which robots can be used in a classroom and some of the advantages of using machines over humans for both students and teachers.

Dimick, Patricia. 2016. "5 Educational Robots You Can Use in Your Classroom." EmergingEdTech. http://www.emergingedtech.com/2016/11/5-educational-robots-for-stem-steam-classroom/. Accessed on October 23, 2017.

> A number of Web sites list "the best" robots available for educational use, of which this is an example.

"Educational Robots." 2017. Robots and Androids. http://www.robots-and-androids.com/educational-robots.html. Accessed on October 20, 2017.

> A host of devices are now available for enhancing the educational experiences of students from the primary grades through college. This Web site provides examples of such robots.

Eisenger, Dale. 2015. "The Mechanical Exoskeleton Shaping the Future of Health Care." *Daily Beast.* https://www.thedaily

beast.com/the-mechanical-exoskeleton-shaping-the-future-of-health-care. Accessed on October 17, 2017.

Eisenger describes some of the specific robotic devices that have been developed for use in rehabilitation programs for individuals with debilitating injuries or illnesses.

Emmi, Luis, et al. 2014. "New Trends in Robotics for Agriculture: Integration and Assessment of a Real Fleet of Robots." *The Scientific World Journal.* vol. 2014, Article ID 404059, 21 pages, 2014. doi:10.1155/2014/404059. Accessed on March 17, 2018.

The authors provide an excellent introduction to the use of robots in agriculture, with some current trends and future prospects in the field. They discuss some technical problems involved in the development of agricultural robots.

Finkelstein, Robert. 2010. "Military Robotics: Malignant Machines or the Path to Peace?" https://www.robotictechnologyinc.com/images/upload/file/Presentation%20Military%20Robotics%20Overview%20Jan%202010.pdf. October 17, 2017.

This very detailed discussion of military robots covers topics such as a brief history of military robots, types of military robots, intelligence and autonomy, state of the technology, military robots for homeland security, commercialization of military robots, and forecasts for the 22nd century.

Fulton, Morven. 2016. "Robotic Process Automation: Friend or Foe for Your Risk Profile." https://www.linkedin.com/pulse/robotic-process-automation-friend-foe-your-risk-profile-morven-fulton/. Accessed on September 24, 2017.

The use of robots in various fields of finance has a number of important benefits that ensure the use of robotic process automation will continue in those fields. This article explores, however, some of the risks that may be involved with the expanding use of robots in finance.

Galliott, Jai. 2017. "Why We Should Welcome 'Killer Robots,' Not Ban Them." The Conversation. https://theconversation.com/why-we-should-welcome-killer-robots-not-ban-them-45321. Accessed on October 17, 2017.

The author points out that we already have "killing machines" similar to robots and that concerns over the future risk posed by robots in warfare may be overblown. For these and other reasons, he says that we should be more open to robots that can be used in military combat.

Goldhill, Olivia. 2016. "Can We Trust Robots to Make Moral Decisions?" Quartz. https://qz.com/653575/can-we-trust-robots-to-make-moral-decisions/. Accessed on October 18, 2017.

In early 2016, a Microsoft chatbot named Tay began to respond to queries with somewhat bizarre comments such as "Repeat after me, Hitler did nothing wrong" and "Bush did 9/11 and Hitler would have done a better job than the monkey we have got now." The company rapidly removed Tay from the Internet and reprogrammed the robot, but the incident raised questions as to how to deal with a machine's ability to make comments relating to value judgments.

Greenmeier, Larry. 2016. "Robot Exoskeletons March in to Link Mind and Body." Scientific American. https://www.scientificamerican.com/article/robot-exoskeletons-march-in-to-link-mind-and-body/. Accessed on October 17, 2017.

The article discusses programs that have been developed to modify military-based robotic programs to civilian applications in the field of rehabilitation.

Gubrud, Mark. 2013. "US Killer Robot Policy: Full Speed Ahead." Bulletin of the Atomic Scientists. https://thebulletin.org/us-killer-robot-policy-full-speed-ahead. Accessed on October 18, 2017.

The author reviews and discusses U.S. Department of Defense directive 3000.09, establishing policies for the development of robots for military use. He suggests that the directive strongly encourages accelerated research on this type of military weapon and what that means for the United States and other nations.

Harvey, Fiona. 2014. "Robot Farmers Are the Future of Agriculture, Says Government." *Guardian.* https://www.theguard ian.com/environment/2014/jan/09/robots-farm-future. Accessed on October 18, 2017.
    This article reviews some of the ways in which robots will be used in agriculture in the future, such as planting, harvesting, and herding livestock.

Heller, Nathan. 2016. "If Animals Have Rights, Should Robots?" *New Yorker.* https://www.newyorker.com/magazine/2016/11/28/ if-animals-have-rights-should-robots. Accessed on October 22, 2017.
    The author attempts to draw comparisons between the animal world and the future world of robots to determine whether the latter should be granted rights similar to those now given to the former.

Hicks, Kristen. 2016. "Robots in Education: What's Here and What's Coming." Edudemic. http://www.edudemic.com/robots-education-whats-coming/. Accessed on October 18, 2017.
    This article provides a good general introduction to the use of robots in education but is especially helpful because of the many links provided to other articles on more specific aspects of the topic.

"Industrial Robots." 2017. IEEE Spectrum. https://spectrum .ieee.org/robotics/industrial-robots. Accessed on October 24, 2017.
    This Web site contains numerous links to articles on special topics within the field of industrial robots, such as snake robots, DNA robots, agricultural applications of robots, robotics contests, fighting robots, and inflatable robots.

Kazmier, Robin. 2017. "Gecko-Inspired Robotic Gripper Could Snag Space Junk." *Nova Now*. http://www.pbs.org/wgbh/nova/next/space/gecko-inspired-robotic-gripper-could-snag-space-junk/. Accessed on October 25, 2017.

> An ongoing problem in space research and space travel is the ever-accumulating amount of "junk" floating in outer space. Finding ways of collecting this junk has, for some time, been a challenging problem for space researchers. NASA engineers have now developed a system to solve this problem. This systems makes use of robots with properties and abilities similar to those of a gecko, an animal that is capable of grabbing and holding onto objects much larger than itself with a relatively small amount of force. This Web site contains a video showing how the device works.

Kleeman, Jenny. 2017. "The Race to Build the World's First Sex Robots." *Guardian*. https://www.theguardian.com/technology/2017/apr/27/race-to-build-world-first-sex-robot. Accessed on August 20, 2017.

> This article discusses research on humanoid robots that can act as companions for humans who may be lonely or incapacitated in one way or another. Such robots may also be capable of sexual interaction. They may also present a number of issues when they are used for these purposes.

Krywko, Jacek. 2016. "Scientists Believe They've Nailed the Combination That Could Help Robots Feel Love." Quartz. https://qz.com/838420/scientists-built-a-robot-that-feels-emotion-and-can-understand-if-you-love-it-or-not/. Accessed on October 22, 2017.

> The author explores the question as to what it means to say that a robot has emotions and reviews current progress in developing machines with such a quality.

Laub, Danica. 2017. "The Road to Reality of Autonomous Vehicles." Penn State. http://news.psu.edu/story/482497/2017/09/

15/public-events/road-reality-autonomous-vehicles. Accessed on October 21, 2017.

This article reports on the Pennsylvania Automated Vehicle Summit, held at State College, Pennsylvania, in September 2017. Some of the questions raised in that meeting were the following: "How will public policy change?" "What will highways, rural roads, and city streets look like?" "How should local, county, and regional governments and planning organizations prepare their infrastructure and other programs?" "How will the introduction of autonomous vehicles affect the economy?" and "How will autonomous vehicles impact traffic fatalities?"

López, Leticia Lafuente. 2016. "Educational Robots at Global Robot Expo." https://elearningindustry.com/educational-robots-global-robot-expo. Accessed on October 18, 2017.

This article is an excellent overview of the many types of educational robots that are currently in use at one location or another around the world, along with some devices that are still in the early stages of development.

Markou, Christopher. 2017. "Robots and AI Could Soon Have Feelings, Hopes and Rights. . . . We Must Prepare for the Reckoning." The Conversation. https://theconversation.com/robots-and-ai-could-soon-have-feelings-hopes-and-rights-we-must-prepare-for-the-reckoning-73462. Accessed on October 22, 2017.

The author notes that as robots become more and more human like, they are likely to take on more human characteristics, such as "feelings, hopes, and rights." He asks, how is such an evolution likely to take place, and how will society deal with these new realities of human-like robots?

Mayer, Jörg. 2017. "Industrial Robots and Inclusive Growth." Vox. http://voxeu.org/article/industrial-robots-and-inclusive-growth. Accessed on October 24, 2017.

The author discusses the possible use of developing nations, how it may affect their economies, and possible effects on labor markets in these nations.

McNally, Phil, and Sohail Inayatullah. "The Rights of Robots: Technology, Culture and Law in the 21st Century." 1988/2001. Kurzweil Accelerating Intelligence. http://www.kurzweilai.net/ the-rights-of-robots-technology-culture-and-law-in-the-21st-century. Accessed on October 22, 2017.

This posting is a reprint of an article originally published in 1988 and republished on this Web site in 2001. It raises a number of fundamental questions as to whether robots have "rights" or not and, if so, what they might be like.

"Military Robots." 2013. All on Robots. http://www.allonrobots .com/military-robots.html. Accessed on October 17, 2017.

This web page provides a good, but brief, overview of the types of robots that are available for use in military situations.

"Military Robots." 2017. Army of Robots. http://www.armyof robots.com/. Accessed on October 17, 2017.

This web page provides a good introduction to many aspects of military robotics, including a history of military robots, some current uses of military robots, and ethics of using military robots, and how military robots work.

"Military Robots." 2017. Robots and Androids. http://www .robots-and-androids.com/military-robots.html. Accessed on October 17, 2017.

This fairly lengthy article describes a number of robots that have been developed for military use, along with photographs of some of those devices. The page also provides links to a wide variety of other robot-related Web sites.

Molyneux, Cándido García, and Rosa Oyarzabal. 2017. "What Is a Robot under EU Law?" Global Policy Watch. https://www

.globalpolicywatch.com/2017/08/what-is-a-robot-under-eu-law/. Accessed on October 22, 2017.

This article reviews the work of an EU project called Robolaw, designed to consider legal changes that will be necessary as a result of the increasing sophistication of intelligent autonomous agents. Some of the issues to be discussed include civil law liability, product safety, autonomous cars and testing, harmonization of technical standards, and safety standards in the health sector. More information about Robolaw is available through the link to that project in the second paragraph of the article.

Mortenson, Eric. 2017. "Autonomous Robots and Drones Will Operate Future Farms." Capital Press. http://www.capitalpress.com/Oregon/20170824/autonomous-robots-and-drones-will-operate-future-farms. Accessed on October 23, 2017.

The author argues that "the 'march from automation to autonomy' will change agriculture forever." He gives a number of examples to support his prediction.

Mouthuy, Pierre-Alexis, and Andrew Carr. 2017. "Growing Tissue Grafts on Humanoid Robots: A Future Strategy in Regenerative Medicine?" *Science Robots.* doi:10.1126/scirobotics.aam5666. https://www.scribd.com/document/341616189/Growing-Tissue-Grafts-on-Humanoid-Robots-A-Future-Strategy-in-Regenerative-Medicine. Accessed on October 23, 2017.

The authors describe ways in which humanoid robots can be designed and constructed as a means of growing musculoskeletal tissue grafts for use as tissue implants. Such devices could be of significant value in the field of regenerative medicine.

Mubin, Omar, and Muneeb Imtiaz Ahmad. 2017. "Robots Likely to Be Used in Classrooms as Learning Tools Not Teachers." The Conversation. https://theconversation.com/robots-likely-to-be-used-in-classrooms-as-learning-tools-not-teachers-66681. Accessed on March 17, 2018.

The authors note that robots can be used in learning situations either as teachers working directly with students or as adjuncts to human teachers. They review the research on this situation and explain why the later practice is likely to become more common in the future.

Newhart, Veronica, and Mark Warschauer. 2017. "How Robots Could Help Chronically Ill Kids Attend School." The Conversation. https://theconversation.com/how-robots-could-help-chronically-ill-kids-attend-school-69197/ Accessed on October 23, 2017.

The authors discuss one of the settings in which robots can be of significant assistance in education: when students are, for one reason or another, unable to be physically present in a classroom. They describe how such systems might work.

Nicol, Will. 2017. "9 Military Robots That Are Totally Terrifying . . . and Oddly Adorable." Digital Trends. https://www.digitaltrends.com/cool-tech/coolest-military-robots/. Accessed on October 17, 2017.

This web page provides illustrations and brief descriptions of some specific robotic weapons that have been developed for military use.

Pethokoukis, James. 2017. "If a Robot Takes Your Job, Should It Pay Your Taxes Too?" The Week. http://theweek.com/articles/681292/robot-takes-job-should-pay-taxes. Accessed on October 23, 2017.

The author has some suggestions about the assessment of tax burdens on robots who have taken over a human job.

Rangarajan, Aravind Krishnaswamy, P. Raja, and Manuel Pérez-Ruiz. 2017. "Task-Based Agricultural Mobile Robots in Arable Farming: A Review." Spanish Journal of Agricultural Research 15(1): doi.org/10.5424/sjar/2017151-9573. https://

www.researchgate.net/publication/314260159_Task-based_
agricultural_mobile_robots_in_arable_farming_A_review.
Accessed on October 23, 2017.

> This article provides an excellent general overview of the
> ways in which robots are used in a variety of agricultural
> operations today.

Renstrom, Joelle. 2015. "Artificial Intelligence. Real Emotion?"
Slate. http://www.slate.com/articles/technology/future_tense/
2015/04/ex_machina_can_robots_artificial_intelligence_have_
emotions.html. Accessed on October 22, 2017.

> This article explores the question as to whether or not
> robots will ever be able to have emotions and, if so, what
> kinds of emotions humans want them to have or will they
> develop as they evolve.

"Robots and Androids." 2017. http://www.robots-and-androids
.com/medical-robots.html. Accessed on October 17, 2017.

> This comprehensive Web site contains information on most
> aspects of robotics, including household robots, robotic
> pets and toys, medical robots, robotic companies, build-
> ing a robot, robot camps, robot fiction, robot games, and
> social robots.

"Robots in Healthcare—Get Ready!" 2017. The Medical Futurist.
http://medicalfuturist.com/robotics-healthcare/. Accessed on
October 17, 2017.

> This article discusses some ways in which robots may be
> used in health care settings in the future, such as surgi-
> cal procedures, phlebotomy, emotional care of the elderly,
> telemedicine, and exoskeletons. Most of the procedures
> discussed are already in use to a greater or lesser extent.

"Salto: Berkeley's Leaping Robot." 2016. YouTube. https://www
.youtube.com/watch?v=xvIk39rkkiU. Accessed on October 25,
2017.

This video shows the actions of a one-legged robot built to mimic the behaviors of the galago, the animal with the greatest ability to leap into the air. The purpose of this research was to learn more about animal kinetics and perhaps to be used in search and rescue or other exploratory operations across surfaces that are otherwise difficult to maneuver. The video is followed by other videos with interesting robots, such as one that mimics the behavior of a bat, a snake robot, and other mini-robots.

Schwartz, Katrina. 2014. "Robots in the Classroom: What Are They Good For?" KQED News. https://ww2.kqed.org/mindshift/2014/05/27/robots-in-the-classroom-what-are-they-good-for/. Accessed on October 18, 2017.

According to the author, some of the uses of robots in the classroom include demonstrating abstract concepts, increasing student engagement, and exposure to computational thinking.

Seasongood, Shawn. 2017. "Not Just for the Assembly Line: A Case for Robotics in Accounting and Finance." FEI. https://www.financialexecutives.org/Topics/Technology/Not-Just-for-the-Assembly-Line-A-Case-for-Robotic.aspx. Accessed on October 18, 2017.

This Web site provides an excellent overview of the many ways in which robots can be, and will be in the future, used for applications in business and finance.

Solon, Olivia. 2016. "The Rise of Robots: Forget Evil AI—The Real Risk Is Far More Insidious." *Guardian*. https://www.theguardian.com/technology/2016/aug/30/rise-of-robots-evil-artificial-intelligence-uc-berkeley. Accessed on August 20, 2017.

The author argues that the risk of robots rising up against humans is negligible but that other risks—such as a robot's making a mistake in its functioning—are far more likely and troublesome.

Stamp, Jimmy. 2017. "Robotics and Fulfillment Centers Are Reshaping Retail—And Cities Could Be Next." *Architects Newspaper*. https://archpaper.com/2017/08/architecture-fulfill ment-centers/. Accessed on October 25, 2017.

> This article discusses in some detail the role that robots play in large fulfillment centers, storage facilities from which online orders are completed. It also comments briefly on ways in which current practices in such centers may have an influence on the structure and function of urban areas in the future.

Thomas, Will. 2014. "Derek Price on Automata, Simulacra, and the Rise of 'Mechanicism.'" Ether Wave Propaganda. https://etherwave.wordpress.com/2014/08/28/derek-price-on-automata-simulacra-and-the-rise-of-mechanicism/. Accessed on October 21, 2017.

> This long article offers a thoughtful review of a historical article by one of the great modern historians of science, Derek de Solla Price, on the role of automata in history. Links to articles by Price and his colleagues are also available within the posting.

"Types of Medical Robots in Healthcare." 2017. http://triotree .com/blog/types-medical-robots-healthcare/. Accessed on October 17, 2017.

> This Web site focuses on a half dozen specific types of robots currently in use in the medical fields, with illustrations for each application.

Van de List, Bobbie. 2016. "Welcome to the Hotel Automata." Discover. http://blogs.discovermagazine.com/crux/2016/12/21/robot-hotel-japan/#.WetqvFtSypq. Accessed on October 21, 2017.

> The writer describes her experience at the first hotel in the world to be staffed and operated entirely by robots. Illustrations support the author's narrative.

Wang, Brian. 2017. "Made in Space Also Working on Robotic Manufacturing of Large Structures in Space." Next Big Future. https://www.nextbigfuture.com/2017/06/made-in-space-also-working-on-robotic-manufacturing-of-large-structures-in-space .html. Accessed on October 25, 2017.

This excellent web page explains how structures can be made in space, a procedure that is likely to become more common in the future. It contains a number of videos that illustrate the processes in action.

West, Darrell M. 2017. "What Happens If Robots Take the Jobs?: The Impact of Emerging Technologies on Employment and Public Policy." Center for Technology Innovation at Brookings. https://www.brookings.edu/wp-content/uploads/2016/06/ robotwork.pdf. Accessed on October 23, 2017.

This is one of the most frequently cited and influential papers on the effects of the growth in robotics on the workforce. It is must reading for anyone interested in this topic.

Introduction

Human research on robotic devices dates back more than 5,000 years. A time line over these five millennia reveals the gradual introduction of more and more complex devices with abilities that approach those of humans. This chapter provides some of the most important of those historical events. The events mentioned are only a small selection of all possible items. Dozens, if not hundreds, of inventors throughout history have produced a panoply of mechanical devices worthy of mention in a chapter such as this one. Space permits the inclusion of only a few examples. For additional references on the history of robotics, see the list of publications at the conclusion of this chronology.

ca. 3500 BCE   Greek legend has it that Hephaestus, the god of blacksmiths, craftsmen, artisans, sculptors, metals, metallurgy, fire, and volcanoes, built the first automata, including a pair of fire-breathing horses, a giant eagle that tortured the god Prometheus, a group of singing women in the shrine of Apollo at Delphi, and a pair of watchdogs at the palace of King Alkinous.

Sofia, a humanoid robot capable of over 62 facial expressions, made her first appearance in India at the Indian Institute of Technology Bombay (IIT-B) during its cultural extravaganza TechFest in Mumbai on December 30, 2017. In October 2017, the robot became a Saudi Arabian citizen, making her the first robot to receive citizenship of any country. (Pratik Chorge/ Hindustan Times)

The first written descriptions of these devices appear in Homer's *Iliad*, written sometime between 700 and 800 BCE.

ca. 420 BCE Greek natural philosopher Archytas invents a mechanical pigeon, capable of flying by means of steam-powered wings. Some historians call this invention the first robot for which there is conclusive proof.

ca. 270 BCE Greek inventor Ctesibius of Alexandria designs a number of mechanical devices, of which the most famous is a water clock. Some of these devices are also human-like and animal-like automata that move and act powered by the use of steam.

ca. 10 BCE The sculpture Pygmalion of Cyprus falls in love with the statue of a woman he has made. Through the actions of the goddess Aphrodite, the statue is brought to life, and she and Pygmalion are eventually married.

ca. 100 CE One of the most famous of inventors of the classical period, Hero of Alexandria, constructs a number of automata, including a pipe organ powered by a windmill, thought to be the earliest device operated by wind power. He also designed and built a mechanical puppet play that lasted nearly 10 minutes in length.

1206 Arabic scholar Ibn Ismail Ibn al-Razzaz Al-Jazari (Al-Jazari) publishes a book, *Kitab Fi Ma Rifat Al-hiyal Al-handasiyya* (*The Book of Ingenious Mechanical Devices*), thought to be the most comprehensive book on automata of the time.

1495 Italian polymath Leonardo da Vinci invents a mechanical knight, which he may or may not actually have constructed during his lifetime.

1500 German inventor Hans Bullmann is credited with building the first true androids, a number of human-like figures that acted under their own power. His musical-instrument-playing figures were widely popular during public gatherings.

1543 While still a student at Cambridge University, John Dee, later Queen Elizabeth I's astrologer, constructs a wooden mechanical beetle that actually flies under steam power. The beetle is used in a student production of Aristophane's "Pax." The beetle is so realistic that some critics believe it is the work of the devil, and charges of sorcery are brought against Dee.

1580 Rabbi Eliyahu Ba'al Shem of Chelm is credited with building the first functioning golem, a figure with a very long history in the Jewish tradition. Reporters of the time say that Shem's golem came to life and performed a number of tasks for him.

1680 Although better known for his contributions to mathematics and physics, Dutch scholar Christiaan Huygens also dabbled in the invention of automata, the most famous of which was a highly accurate mechanical clock. In 1680, he was also instructed by the king of France to make for his enjoyment a large collection of mechanical figures to be used in mock military battles.

1725 German inventor Lorenz Rosenegge constructs a complex, miniaturized, automated reproduction of a village consisting of 256 figures, of whom 119 are automated with the use of water power. The village is one of the ultimate achievements of an art developed by many inventors of the preceding decades who have constructed small communities of individuals, animals, buildings, and other structures, most commonly powered by water.

1739 French artist and inventor Jacques de Vaucanson builds a mechanical duck that carries out many of the behaviors associated with live ducks, such as sitting, standing, splashing around in water, quacking, eating and digesting food, and eliminating wastes. Another of his earlier inventions, a flying angel, resulted in his being dismissed from the Jesuit order because of suspicions about a "dark influence" on his work.

1753–1760 Austrian inventor Friedrich von Knaus (also Knauss) invents a series of automata featuring a human figure writing at a table, apparently the first example of automated writing machines. The final design stood nearly eight feet tall, and the automaton writer was capable of producing a text of up to 107 words.

1772   German watchmaker, Peter Kinzing, and cabinet-maker, David Roentgen, invent the Joueuse de Tympanon, an approximately 20-inch female automaton that plays the tympanon, a dulcimer-like stringed instrument. The figure is presented to Queen Marie Antoinette of France in 1784.

1773   Swiss inventors Pierre and Henry Louis Jaquet-Droz produce a series of animated human figures capable of carrying out a number of functions, such as writing, drawing, and playing music.

1818   English novelist, Mary Shelley, publishes *Frankenstein*, the story of a monster created from the nonliving remains of dead humans.

1820s   Japanese engineer and inventor Tanaka Hisashige begins the production of small *karakuri-ningyo* dolls, mechanized puppets capable of performing a variety of relatively simple tasks, such as accepting an empty tea cup and returning it full to a person, shooting arrows, and writing calligraphic symbols. The dolls become widely popular, and Tanaka travels around the country demonstrating his puppets before noble families and at regional fairs.

1890   Thomas Edison makes available for sale a "talking doll" that contains a wax cylinder that allows the doll to "speak" for a period of six seconds. The doll was a commercial failure because its voice was so harsh and raspy as to frighten beyond endurance children to whom it was given as a toy.

1893   Canadian inventor George Moore builds a "steam man" that is able to walk on its own, smokes a cigar, and looks at its surroundings. It is powered by a steam engine inserted into its body.

1900   American inventor Phillip Louis Perew invents an "electric man" somewhat similar to Moore's steam man. The electric man appears to walk like a human, pulling a cart behind him.

In fact, the drive train for the system is stored in the cart, which drives the electric man ahead of it. In its last version, the system is powered not by an electric motor but by a gasoline engine.

1921   Czech playwright Karel Čapek introduces the term *robot* in his play *R.U.R.* to describe androids that eventually take over the world. The title is an acronym for the phrase *Rossumovi Univerzální Roboti* (*Rossum's Universal Robots*).

1926   Ohio State University psychologist Sidney L. Pressey describes the earliest type of "teaching machine," a robot-like device through which a learner can instruct himself or herself in some specific subject area.

1927   German film director Fritz Lang releases the motion picture *Metropolis*, one of whose characters is the first gynoid to appear in films.

1927   Westinghouse inventor Roy Wensley builds Herbert Televox, a robot designed to answer and respond to telephone calls. The device works by translating various sound pitches, each of which is programmed to produce some specific mechanical response. A detailed description of the robot's operation can be found at http://drloihjournal.blogspot.com/2016/11/herbert-televox-mechanical-man-chicago.html.

1928   At the Kyoto Grand Exposition in Commemoration of the Imperial Coronation, Japanese biologist and editorial writer Nishimura Makoto unveils *Gakutensoku*, generally thought to be the first robot built in Japan. The robot's name means "learning from nature" and is intended to represent the harmony of nature. It was able to produce a variety of facial expressions and write programmed messages with a mechanical pencil.

1934   American inventors Willard Pollard and Harold Roselund file a patent application for a robotic paint-spraying machine. Some controversy remains as to whether or not this device is the world's first spray-painting robot or if that accolade falls to a similar invention put into use in Norway in 1969 (q.v.).

1942   In his short story "Runaround," writer Isaac Asimov introduces the Three Laws of Robotics, created to limit the extent to which robots could overtake human activities. The Three Laws have been the subject of considerable discussion and debate ever since and have undergone a number of modifications and revisions. They have become controversial especially in the past few decades, as the possibility of more life-like androids and gynoids has been created.

1948–1949   English neurophysiologist William Grey Walter builds the first social robots. *Social robots* are defined as robots that interact with humans and/or other objects in their environment. Walter's research was motivated by his belief that relatively simple connections between a limited number of neurons can result in complex human behavior. In his robots, Elmer and Elsie (ELectroMEchanical Robot, Light-Sensitive), neurons and their connections were modeled by simple electric circuits. The robots are sometimes known as Walter's turtles or Walter's tortoises because of their physical appearance: a shell-like structure that operated on three wheels at a relatively slow speed.

1950   British mathematician Alan Turing devises a test designed to distinguish between a human and a robot. The test depends on an independent observer's ability to tell whether he or she is interacting with the unknown robot/human kept out of sight of the observer. A number of modifications of the Turing test have been devised over the past seven decades, as researchers attempt to find ways of deciding what constitutes robot behavior and what does not.

1954   Harvard psychologist B. F. Skinner designs perhaps the best known of early "teaching machines," robotic-type devices by which individuals can pursue their own program of self-directed learning.

1958   The U.S. National Space and Aeronautics Administration (NASA) initiates its Ranger spacecraft program, designed to provide images of the moon's surface in preparation for the landing of the first humans on the moon. The first six launches ended in failure, but Rangers 7, 8, and 9 all sent back usable photographs before crashing on the moon's surface.

1961 Stanford University graduate student James L. Adams constructs the first in a long series of *Stanford Carts*, devices designed to study the movement of robots through the use of television cameras and computer-designed instructions. Such devices used the TV camera to provide images of the cart's surroundings, which were then relayed to the computer. The computer next determined the proper path for the device to follow in order to avoid collisions with surrounding objects. The Stanford Cart was seen as the prototype of mechanical devices that could be used for exploration of the moon and Mars.

1961 The first mechanical robot, called Unimate, is developed by American inventor George Devol. The robot is first placed into operation on the assembly line at General Motors plant in Ewing Township, New Jersey. It is designed to transport die castings used in the production of automobile bodies.

1964 General Motors orders 66 Unimates to be used in its assembly plant at Lordstown, Ohio. The robots are used for welding, converting the process to 90 percent robot-controlled at the plant.

1965 GE engineer Ralph Mosher develops the so-called cybernetic anthropomorphous machine (or CAM) for the U.S. Army. The robot was intended to carry heavy equipment for infantry forces over rough terrain. Movement of the device (at a speed of about five miles per hour) was controlled by an onboard operator.

1966 The first robot in the NASA Surveyor space program lands on the moon. The seven spacecraft in the Surveyor program were all equipped with television cameras. In addition, some robots carried mechanical arms for collecting soil samples and/or devices for determining the chemical composition of the soil.

1967 The first mechanical robot in Europe, a Unimate (*see* **1961**), is put into operation at the Metallverken assembly plant in Upplands Väsby, Sweden.

1968 MIT physicist Marvin Minsky invents a crab-like robot with 12 arms for underwater operations.

1968 Robert McGhee and graduate student Andrew Frank design the Phony Pony (also known as the California Horse), the first walking machine whose actions are controlled by an installed computer.

1969 Researchers at Stanford University's SRI's Artificial Intelligence Center invent Shakey, the first robot with the ability to perceive and reason about its surroundings.

1969 Norwegian inventor Ole Molaug introduces the first spray-painting robot, first used to spray enameling on bath tubs (*also see* **1934**, Pollard and Roselund).

1969 Researchers at the University of Edinburgh begin a program to develop a robot that makes use of artificial intelligence. The first device produced was called Freddy (or Freddy I or Freddy Mark I), although the most famous product of the research was a later model, Freddy II. Robots in the series were capable of viewing a scene presented to them before carrying out one or more operations on the scene (e.g., stacking blocks). Research of this type continues today at the university's Institute of Perception, Action and Behaviour.

1970 Hitachi corporation introduces a vision-based robot capable of reading blueprints and constructing an object based on those plans.

1970 The Soviet Space Agency launches Lunokhod 1, a robot designed for exploration of the moon's surface. The device lands on the surface of the moon on November 17 and continues to send back data to Earth until it is deactivated by controllers on September 14, 1971.

1971 The world's first national robotic association is formed in Japan as the Industrial Robot Conversazione, later to become the Japan Robot Association in 1973.

1973 The Stanford arm, designed by Stanford mechanical engineering student Victor Scheinman, is placed into operation as a mechanism for augmenting the movements of a human arm. It is capable of recognizing a target, planning for steps in an operation, quantifying the force on an object, and manipulating objects by means of some predetermined pattern.

1973 The German (now Chinese-owned) KUKA manufacturing company releases a new type of robot consisting of six independently operating arms, the first in a series of Famulus robots.

1973 Inventor Richard Hohn develops a robot called The Tomorrow Tool or T3, whose actions can be controlled by instructions from a minicomputer.

1974 Allmänna Svenska Elektriska Aktiebolaget (General Swedish Electric Company; ASEA) releases for sale a robot whose actions are controlled by an 8-bit microprocessor with a memory capacity of 16 kb and a human-like arm capable of lifting 6 kg around a range of five axes.

1975 Olivetti company produces one of the first Cartesian-coordinate robots, Sigma, for use in assembly lines. A cartesian-coordinate robot is one whose arms move in linear, rather than rotational, directions.

1979 Carnegie Mellon professors Raj Reddy and Angel Jordan, along with Tom Murrin, president of the Westinghouse Electric Corporation, establish the Robotics Institute at the university, with the aim of "dream of ushering in a new age of thinking robots."

1981 A research team led by Mark Raibert at the Carnegie Mellon Robotics Institute designs a one-legged robot capable of demonstrating and studying principles of dynamic balance. Two forms of the invention are called Thumper and Bow Leg Hopper.

1981 Japanese inventor Hirose Shigeo designs the original Titan walking robot. He continues with at least 10 improved models that are eventually able to transverse both level and uphill ground, as well as steps. The Titan series is of special interest because its robots are large enough to carry small loads, but not so big as to be unwieldy. (For an image of a later Titan model, see http://menzelphoto.photoshelter.com/image/I0000h31uUJ2zW1A.)

1982   Healthkit company releases the first of a series of HEROs (Healthkit Education Robots). Although capable of carrying out a few elementary domestic tasks, HEROs were designed primarily for educational and entertainment purposes. An estimated 21,000 units of three models of the HEROs were sold in the 1980s. Today, the robots are considered relatively rare and collectors' items.

1983   A research team led by William Whittaker designs two robots for exploring and cleaning up the basement of the damaged Three Mile Island nuclear reactor near Harrisburg, Pennsylvania. The robots are outfitted with cameras, lights, radiation detectors, vacuums, scoops, scrapers, drills, and a high-pressure spray nozzle.

1983   A surgical robot, Arthrobot, is the first such device used for medical purposes. It is designed to assist surgeons in orthopedic procedures. It first use occurs in Vancouver, Canada.

1984   The Marine Technology Department of the Japan Marine and Science Technology Centre initiates research on underwater octabots (eight-legged robots) designed to study the ocean floor. Three models of the robot are eventually produced and tested. Modern octabots are sold for more mundane purposes, such as cleaning homes and commercial swimming pools. (More details are available at http://cyberneticzoo.com/underwater-robotics/1985-aquarobot-aquatic-walking-robot-japanese/.)

1985   The first nonorthopedic robotic surgery is performed using the PUMA (Programmable Universal Machine for Assembly, or Programmable Universal Manipulation Arm) robot for the insertion of a needle into the brain in a biopsy procedure. The procedure is not yet approved for general use, since PUMA had been developed for and used strictly in industrial procedures prior to this time.

1989 MIT researcher Rodney A. Brooks designs a six-legged robot (hexapod) intended to perform many of the same functions as those exhibited by insects and other six-legged creatures. (Video of a more recent version of hexapod is available at https://www.youtube.com/watch?v=hjaKstJaa9Y.)

1992 A robotic system called ROBODOC is used in conjunction with a human surgeon to perform a total hip arthroplasty. The procedure is not formally authorized by the U.S. Food and Drug Administration (FDA) until August 2008.

1994 The FDA issues the first approval for robotic surgery. The robot, AESOP (Automated Endoscopic System for Optimal Positioning), is used for endoscopic surgery in which a surgeon's hand and arm movements are mimicked with maximum accuracy by the robot.

1995 The Japanese electronics firm NEC produces a snake-like robot called Quake Snake designed to search otherwise inaccessible locations for trapped individuals, such as buried regions following an earthquake or explosion. Such devices, also called *snakebots*, contain a small television camera at its head, allowing an operator to direct the snake's movement. Later versions of the robot have been used following disasters such as the Niigata Chuetsu earthquake in Japan in 2004 and the Fukushima earthquake and tsunami of 2011. (For a photograph of an early snakebot, see https://www.researchgate.net/figure/243776586_fig5_Figure-2-13-The-NEC-%27Quake-Snake%27-utilized-a-novel-universal-type-joint-between-links.)

1996 Electrolux, the Swedish corporation, demonstrates the first fully automated domestic vacuuming system, called the Trilobite. The product becomes commercially available in 2001.

1996 Computer Motion corporation makes available an adaptation of AESOP (*see* **1994**) that understands and performs procedures from a surgeon's oral commands.

1996   The MIT Department of Ocean Engineering produces a robot called Robotuna to simulate the mechanisms by which fish swim. It choose the fastest of all fish, tuna, around which to name the project. The goal of the project is not only to better understand swimming mechanisms in fish but also to develop the fastest and more efficient systems for ship propulsion. (A later model of the robot can be seen at https://www.you tube.com/watch?v=w5-6F8RknUM.)

1997   MIT researcher Cynthia Breazeal begins work on the development of a social robot, which she eventually calls Kismet. The robot can speak, display a variety of facial expressions, look in different directions, tilt and swivel its head, understand and respond to human speech, and carry out other human-like actions.

1998   iRobot Corporation produces the first packbot, a multipurpose device designed originally for military use. It can be used for bomb disposal, surveillance, and reconnaissance. In addition to military operations, such as its use in Iraq and Afghanistan, it was eventually used to search the remains of Twin Towers' collapse on September 11, 2001, and in the investigation of the damaged Fukushima nuclear reactor in 2011.

1998   The American electronics firm Tiger Electronics introduces the world's first robotic pet, an owl-like creature named Furby. Furbies originally speak their own language but are able to learn other languages. They achieved enormous commercial success, with sales of more than 40 million units over the three years during which they were available. Furbies were later reissued in a variety of adaptations, such as Furby Babies, Furby Friends, Emoto-Tonic Furby Friends and Babies, and a Furbacca model.

2000   The FDA approves the use of the da Vinci surgical system, developed by Intuitive Surgical Inc. for laparoscopic procedures. The device makes use of three-dimensional vision, surgeon-assisted "fingers," closely controlled motion devices, ergonomic balance, and precision superior to that typical of human actions. As of early 2018, da Vinci is one of the most widely used surgical robots in the world, with applications extending far beyond the original laproscopic procedures.

2000 The Singularity Institute for Artificial Intelligence is created to study social, ethical, economic, and other issues evolving as a result of research on artificial intelligence. The institute is now known as the Machine Intelligence Research Institute.

2000 Japan's Honda corporation releases the product of a five-year research program known as ASIMO (Advanced Step in Innovative Mobility). The android has an appearance and behavior close to that of humans and is capable of walking on its own, following a person, going in a direction indicated by the person, approaching and greeting someone, and recognizing and responding to a human. ASIMO later goes through four major modifications and improvements.

2000 The first robotic prostatectomy (prostate removal) is performed by a surgical team led by German surgeon Jochen Binder. The procedure is currently the most common type of robotic surgery performed in the United States.

2001 The Space Station Remote Manipulator System (SSRMS) goes into operation. The device, built by MDA Space Missions for the Canadian Space Agency, consists of three parts: the Space Station Remote Manipulator System (SSRMS), more commonly known as Canadarm2; the Mobile Remote Servicer Base System (MBS); and the Special Purpose Dexterous Manipulator (SPDM), also known as Canada hand or Dextre.

2002 iRobot manufacturing company, after a decade of producing a variety of military robots, releases its first domestic product, the Roomba home vacuuming system. By some estimates, a million copies of the device had been sold by 2004.

2003 NASA sends two robots to Mars to study the planet's atmosphere and surface, Opportunity and Spirit. Although designed to survive for only about three months, both rovers continued working far beyond that deadline, Spirit to 2011 and Opportunity to (as of early 2018) the current date.

2005   Researchers at Cornell University build the first self-replicating robot, a device that is capable of making an exact copy of itself. The Cornell robot has no useful function other than demonstrating the possibility of building such a device. For an illustration of the robot in action, see https://www.you tube.com/watch?v=K_EWzxRn8Xo.

2005   With research funding from DARPA (Defense Advances Research Projects Agency), the firm of Boston Dynamics develops Big Dog, a quadrupedal robot able to run at speeds of 10 km/h (6 mph), climb slopes of up to 35 degrees, walk across rubble, climb muddy hiking trails, walk in snow and water, and carry loads of up to 150 kg (300 lb).

2005   Murata Manufacturing Company releases Murata Boy, a self-balancing bicycle-riding robot that remains upright even when the bicycle is not in motion (*also see* **2008**).

2006   RoboCoaster company installs the first robot designed for entertainment parks at the Futuroscope park in Poitou-Charentes, France. As of 2017, a dozen more such robots have been installed in eight countries, China, Denmark, France, Germany, Japan, United Arab Emirates, the United Kingdom, and the United States. (For video of an early model of the device, see https://www.youtube.com/watch?v=XjSN4fWemxE.)

2006   Caleb Chung, inventor of Furby robots (*see* **1998**), introduces Pleo, a robotic dinosaur capable of responding to and reacting with a number of human commands and actions.

2008   The Japanese firm Murata Manufacturing releases Murata Girl, a self-balancing, unicycle riding robot (also see **2005**).

2009   Boston Dynamics corporation unveils a robot claimed to be the world's first anthropomorphic robot, Petman (Protection Ensemble Test Mannequin). Funded by DARPA, the project is originally intended to develop protective gear for individuals sent into hazardous environments. The project is the forerunner of Atlas (*see* **2013**), q.v.

2011   NASA launches the first humanoid robot (Robonaut) into space. It is sent to the International Space Station, where it will assist astronauts with a variety of routine tasks. Originally mounted on a pedestal in the station, it will eventually be provided with legs that will allow it to move throughout the station much as humans do.

2012   Nevada becomes the first state to issue driver's licenses for self-driving cars.

2013   Boston Dynamic produces its most recent version of anthropomorphic robots, Atlas. The robot is funded by DARPA for the development of human-like machines capable of conducting hazardous search and rescue missions. (For the most recent [2016] version of the robot, see https://www.you tube.com/watch?v=rVlhMGQgDkY.)

2014   The Future of Humanity Institute is formed at Oxford University for the purpose of studying possible future threats to the existence and viability of human civilization. Among the research topics at the institute is the development of robotics and artificial intelligence.

2015   A robot built by Yale researchers Brian Scassellati and Justin Hart, Nico, passes a "self-awareness" test by recognizing itself in a mirror. The experiment is the first successful demonstration relating to a robot's self-consciousness. (A demonstration of the experiment is available at https://www.youtube .com/watch?v=jx6kg0ZfhAI.)

2015   A group of internationally eminent scientists, including Stephen Hawking, Martin Rees, Elon Musk, Bill Gates, and Steve Wozniak, publish an open letter, "Research Priorities for Robust and Beneficial Artificial Intelligence," in which they comment on progress in robotics and artificial intelligence and its promises for the improvement of mankind. It also warns of pitfalls to human civilization inherent in the development of more intelligent machines.

2016   A research team under the direction of Kit Kevin Parker at Harvard University produces the world's first artificial animal, a hybrid consisting of a soft elastomer body with embedded flakes of gold and a superimposed collection of rat heart cells. The device responds with an undulating movement when exposed to light. Researchers can use the device for research on biological actions, with the possible development of more complex replacement body parts and more advanced hybrid animals.

2016   American engineer and inventor Dean L. Kamen founds FIRST (For Inspiration and Recognition of Science and Technology) Global, a nonprofit organization designed to promote STEM (science, technology, engineering, and mathematics) education. The organization sponsors an annual Olympics-style robotics competitions for young adults called the FIRST Global Challenge.

2016   Google Brain releases a new set of Robotic Laws, intended to replace those first proposed by Isaac Asimov in 1942. (The laws are presented and discussed at https://www.fastcode sign.com/3061230/google-created-its-own-laws-of-robotics.)

2017   The European Parliament rejects a plan to create a tax on companies whose robots displace human workers but calls for member nations to begin developing laws and regulations relating to the development and implementation of robots in their workforces.

2017   Stephen Hawking produces a new documentary film *The Search for a New Earth*, in which he warns that humans have less than 100 years to find a new place in space to settle and continue civilization. The warning is based on his predictions about the replacement of humans on Earth by robotic devices.

2017   The appearance of Hanson Robotics' new interactive robot, Sophia, on late-night television reinforces concerns about risks of ongoing developments in robot science when she announces that she has plans to "dominate the human race."

Hanson quickly develops a patch to replace this line of "thinking" in the robot.

2017   A research team at Gallaudet University produces the prototype for a new robot capable of teaching American Sign Language and other languages to babies as young as six months. The device, called RAVE (for Robot Avatar Thermal-Enhanced prototype), is able to sense the emotional state of babies and determine the point at which they are ready for language learning.

Additional References

Dinwiddie, Keith. 2016. *Basic Robotics*. Boston: Cengage Learning, Chapter 1.

"History of Robots." 2008. https://wiki.nus.edu.sg/display/cs1105groupreports/History+of+robots. Accessed on October 12, 2017.

"Robot History." 2017. International Federation of Robotics. https://ifr.org/robot-history. Accessed on October 12, 2017.

Robotic Timelin. 2012. "Plastic Pals." http://www.plasticpals.com/?page_id=26736. Accessed on October 12, 2017.

Rose, Simon, and Avi Abrams. 2015. "Dark Roasted Blend." http://www.darkroastedblend.com/2015/01/amazing-automatons-robots-victorian.html. Accessed on October 12, 2017.

"Timeline of Robotics." 2014. The History of Computing Project. https://www.thocp.net/reference/robotics/robotics.html. Accessed on October 12, 2017.

Discussions of robots and robotics often involve terminology that is unfamiliar to the average person. In some cases, the terms are scientific or technological expressions used most commonly by professionals in the field. In other cases, the terms may be part of the everyday vernacular that some people may *think* they understand but that actually have more precise meanings. This glossary defines some of those terms that have been used in this book, along with some terms that one may encounter in additional research on the topic.

actuator   A motor that converts the energy of some external source into mechanical movements by a robot.

aerobot   A robot capable of independent flight.

android   A robot that looks and/or acts like a male human.

articulated arm coordinate robot   A robot whose arms consist of discrete segments connected by mechanical joints.

articulated manipulator   A robotic arm that consists of sections joined to each other, allowing the arm to move in various directions at different positions on the arm.

artificial intelligence (AI)   The branch of mathematics that deals with intelligent behavior by machines.

artificial intelligence singularity (AI singularity)   A point in time when the growth of robotic intelligence begins to increase so rapidly that humans are unable to control its further development or the consequences it may have on human society.

assembly robot   A type of industrial robot designed specifically to bring and fit together component parts of some complex device.

augmentation   A medical program designed to help an individual to adjust to the loss of some body part or basic physical function.

automaton   (plural: automata or automatons) A mechanism made of inanimate materials that performs human-like tasks by following some set of instructions provided by a human.

autonomous vehicle   A vehicle capable of driving itself, without human guidance; also known as **self-driving vehicle**.

bot   An abbreviation for *robot*.

CAM.   *See* **pedipulator**.

chatbot   A robotic program designed to carry on conversations with a human, especially over the Internet.

comfort robot   Some type of robot designed to be used by humans to deal with a social or emotional deficiency in their lives.

contact sensor   A device by which a robot is able to detect some exterior stimulus, such as pressure or light.

control command   An instruction provided to a robot by a human.

cybernetic anthropomorphous machine (CAM).   *See* **pedipulator**.

cyborg   A device that consists of both human and mechanical parts; also known as a cybernetic organism.

cylindrical-coordinate robot   A robot whose arms move in a cylindrical motion out of a central axis.

degrees of freedom   The number of directions in which a robot can move, based on a three-dimensional axis of Cartesian coordinates.

feedback system   A series of actions by which a robot becomes aware of the consequences of its actions and adjusts its behavior in some way to change those consequences.

fembot *See* **gynoid**.

flexicurity (flexible security) A social system that provides financial security for an individual whether or not he or she is actually employed.

gantry robot A robot that moves along a horizontal beam.

golem A legendary type of automaton, usually made out of clay, created to serve its master in some way or another.

gripper. *See* **manipulator**.

gynoid A humanoid robot designed to look like a human female; also known as a *fembot*.

hexapod A robot with six legs.

humanoid A robot designed to either look like or act like a human, or both.

hydraulics The field of physics that involves the transfer of energy by much of a liquid, usually water. The movement of automata and other simple robots is often powered by hydraulics.

individual activity account (IAA) An economic program that allows a person to set aside a certain amount of money in a specific time period as a savings account for retirement or future needs.

industrial robot A robot designed to carry out some industrial process, such as moving objects, riveting, spraying, or shaping materials.

manipulator (also **gripper**) A robot "hand" designed to pick up, hold, and transfer objects.

octapod A robot with eight legs.

pedipulator A machine capable of walking like a human, or the "legs" of such a machine.

pendant (teach box) A control box by which programming commands are sent to a robot.

pneumatics The field of physics that deals with the movement of objects by means of differences in gas (usually air) pressure. Many automata and simple robots carried out their functions on the basis of pneumatic principles.

proportionality   A military concept that limits the damage done to civilian populations based on their concomitant military benefits.

rectangular-coordinate robot   A robot whose arms move in straight lines along the x, y, and z axes of space.

rehabilitation   A medical program designed to assist a person in recovering some bodily function(s) damaged because of injury or disease.

robonaut   A robot designed to carry out its functions in some form of space travel.

robot camp   An extracurricular program designed, in most cases, for precollege students and intended to teach the basic science of robotics.

robotics   The field of technology that deals with the design, construction, operation, and uses of robots.

rotary joint   A robotic joint that acts in a circular fashion.

SCARA (Selective Compliance Assembly [or Selective Compliance] Robot Arm)   A robot whose arms move in two Cartesian directions but not in the third direction.

self-driving vehicles   *See* **autonomous vehicle**.

service robot   A robot designed to carry out some type of domestic or other nonindustrial function, such as vacuuming or cleaning an outdoor pool.

servomechanism   A procedure for controlling the movement of a robot. The procedure begins when a controller issues a command to a robot, whose effect is measured by internal sensors. The sensor then sends to the controller the degree to which the command has produced the desirable result. The system continues until the robot has achieved the final position or action desired.

simple machine   A device used to change the direction or amount of force on an object. The six basic simple machines are the lever, wheel and axle, pulley, inclined plane, wedge, and screw.

snake robot   A robot with the physical appearance of a snake, capable of and designed for the purpose of going into otherwise inaccessible spaces looking for humans or other objects.

social robot   A type of robot designed to understand and respond to communicate and interact with other humans or other robots.

spherical-coordinate robot   A robot whose arms move in all three directions from some central pivot point.

surgical robot   A robot designed to carry out or assist a surgeon with some type of surgical procedure.

teach box   *See* **pendant**.

telemedical system   A system by which participants in a medical system (doctor with doctor or patient with doctor) are able to communicate with each other over significant physical distances.

universal basic income   An economic program that ensures that every person receives, by some manner or another, a basic annual income that allows him or her to live a safe and healthy life.

A. I. Artificial Intelligence (film), 37
ABB Robotics, 186
accountability, robotic warfare and, 105
actuator, 61
Adam (specialized robot), 137
Affordable Care Act, 73
agricultural robots, 86–87
Aguirre, Anthony, 179
Aiken, Henry, 19
Alda, Alan, 179
Al-Jami 'bayn al-'ilm wa 'l-'amal al-nafi 'fi sina 'at al-hiyal (A Compendium on the Theory and Useful Practice of the Mechanical Arts) (al-Jazari), 188–189
al-Jazari, 8–9, 188–190
Allen, Colin, 77, 80–81
Allen, Paul, 67
AlphaGo, 31
Amazon Motion to Quash a Search Warrant, 251–253

American Association for Artificial Intelligence, 166
amoral acts, 78–79
androids
    artificial intelligence and, 27
    defined, 23
    features of, 24
Archimedes, 10
Archytas of Tarentum, 6
Argonautica (poem), 35
"Arlequinada" (ballet), 17
artificial intelligence (AI), 27
    critics of, 65–67
    FHI and risks of, 178
    FLI and myths about, 179
    humanoid robots and, 30–31
    Minsky and, 198–199
    Musk on, 63
    strong, 65
    Turing test and, 31
artificial intelligence singularity (AI singularity), 63–65

Asimov, Isaac, 36, 61–62, 82,
    163–166
Asociación Española de
    Robótica, 186
assembly robot, 173, 203
Association for the
    Advancement of Artificial
    Intelligence (AAAI), 65,
    166–167
Associazione Italiana di
    Robotica e Automazione,
    186
augmentation, 96
Automata (Heron), 7
"Automata" (Wright), 36
automata/automaton, 4–20
    in ballets, 16–17
    in Bible, 4–5
    in China, 6
    components of, 9–11
    defined, 4
    as demonstrations of
        magical processes, 12
    fall of Rome and knowledge
        about, 7
    from, to robots, 18–20
    golden age of, 1860–1910,
        14–17
    golem as earliest, 4–5
    in Greek civilization, 5,
        6–7
    in Indian civilization, 5–6
    as instructional devices, 12,
        13–14
    inventing/building, reasons
        for, 12–14

Islamic civilization and,
    7–9
knowledge of, transmitted
    from Islam to Europe, 11
in literature, 17
in modern world, 17–18
in religious ceremonies,
    12–13
Talos as, 5
as toys, 12, 13
autonomous care systems,
    144–145
autonomous vehicle,
    231–232

Babbitt, Seward, 19
Barthelmess, Ulrike, 82
Baum, Frank, 35
Bayt-al-Hikma (House of
    Wisdom), 8
bhuta vahana yanta (royal
    mechanical robots), 5–6
Big Dog, 102
biotechnology, FLI and, 179
Blade Runner (film), 37
Blue Ocean Robotics, 186
Boström, Nick, 167–169,
    177, 179, 184
bot, 18
bottom-up robotics, 169–170
Breazeal, Cynthia, 80
British Automation &
    Robotics Association, 186
Brooks, Rodney, 169–171
business/finance robots,
    87–89

Campaign to Stop Killer
  Robots, 103–106
Čapek, Karel, 20–21,
  36, 130
care robots, 144–147
Caus, Jean Salomon de, 16
chatbots, 156–159, 190
China Robot Industry
  Alliance, 186
Chita-Tegmark, Meia, 179
Christiansen, Ole Kirk, 108
Clever Dummy, A (silent
  film), 36
Cleverbots, 157, 159
CLOOS, 186
Collaborative Robots
  Whitepaper, 204
comfort robots, 92–94
computer-assisted surgical
  systems, 228–229
"Coppélia" (ballet), 17, 35
criminal robots, 247–249
Ctesibius of Alexandria, 6,
  10–11
CyberKnife robotics, 138
cybernetic
  anthropomorphous
  machine (CAM), 26
Cybernetic Zoo
  (Web site), 29
cyborg, defined, 24

da Vinci, Leonardo, 15,
  28–29, 193–195
Da Vinci Surgical
  System, 96

efficacy of, 2013, 224–226
Daihen, 186
Danish Industrial Robot
  Association, 186
de Vaucanson, Jacques,
  211–213
Deep Blue, 30–31
deep neural networks
  (DNN), 38
degrees of freedom, 23, 150
Delibes, Léo, 35
delivery system robotics,
  102
Della Tour, Gianello, 16
Der Sandmann
  (The Sandman)
  (Hoffmann), 17
Devol, George, 24, 171–174
Die Puppe (The Doll)
  (Hoffmann), 17
"Die Puppenfee"
  (ballet), 17
disruptive technologies,
  68–71
domestic robots, 39–40
Draft Report (on Robotics)
  of the European
  Parliament, 2016,
  233–236
drones, 41
  Idaho Drone Law (2013),
    226–228
  morality and, 77
  Perdix system, 101
Duck (de Vaucanson
  automaton), 212

earned income tax credit
  (EITC/EIC), 72
Edison, Thomas, 19–20
Edmonds, Dean, 198
education
  in jobless future, 74–76
  robots and, 90–92
Efficacy of Robotic Surgery,
  2013, 224–226
Electro (humanoid robot),
  32–33
ELIZA (chatbot), 157
Eliza Effect, 157
Engelberger, Joseph, 173,
  174–177
Enigma code, 208
entertainment robots,
  42–43
Eric (Great Britain humanoid
  robot), 32
essays
  Hooper, Rich, 129–132
  Johnson, David E.,
    132–136
  Okpechi, Samuel C.,
    136–140
  Repp, Sierra, 140–144
  Rismani, Shalaleh,
    144–147
  Sarkar, Anjali A., 147–151
  Vadiee, Nader, 151–156
  Zimmerman, Erin,
    156–161
Etzioni, Oren, 82–83, 201
Eugene Goostman (Russian
  machine), 31

Ex Machina (film), 24, 37
exoskeleton, 97–98
"Expedition New Earth"
  (documentary film), 108

Faber, Joseph, 16
Fallon, Jimmy, 64
Fan Hui, 31
feedback system, 96, 148
fembot, 24. *See also* gynoids
FHI. *See* Future of Humanity
  Institute
Field Robotics Center, 205
fire-fighting devices,
  102–103
FIRST (For Inspiration and
  Recognition of Science and
  Technology) Robotics, 43,
  109, 141
FIRST LEGO League (FLL)
  competitions, 134–135
FLI. *See* Future of Life
  Institute
Florida Law on Driverless
  Vehicles, 2016, 231–232
Flowers, Woodie, 109
Flute Player, 211
Forbidden Planet (film), 37
Freddy (robot), 27
Freeman, Morgan, 179
Friendly AI, 196
full moral agency, 80, 81
functional morality, 80
Furbach, Ulrich, 82
Future of Humanity Institute
  (FHI), 65, 177–178

Future of Life Institute (FLI),
65, 178–180
Futureworld (film), 37

Gates, Bill, 72
General Electric
Company, 26
Goalkeeper CIWS (weapon
system), 42
golden age of automata,
1860–1910, 14–17
golem, 4–5, 130
Goliath, The, 100
Google Brain, 83
Guidelines for Robotic Safety
(OSHA), 218–224
accidents, 218–219
hazards, 220–222
personnel, control and
safeguarding,
222–224
robotic safeguarding
system, 219–220
guilty mind, 248
gynoids
artificial intelligence
and, 27
as comfort robot, 92–93
defined, 23–24
features of, 24

Hammond, John, Jr., 20
Harmony (gynoid sex robot),
92–93
Hawking, Stephen, 62–63,
108, 179, 200

Hawkins, Jeff, 66
health care robots, 94–99
Okpechi essay on,
136–140
Hephaistos (Hephaestus), 5
Hermes Trismegistus, 5
Heron (Hero) of
Alexandria, 3, 6, 7, 13,
180–181
Hill, Wycliffe, 32
Hiroshi Ishiguro, 34,
187–188
Hoffmann, E.T.A., 17
Honda Company, 34–35
Hooper, Rich, 129–132
Hughes, James J., 184
human society, future of
robots in, 61–109
agricultural tasks and,
86–87
business/finance and,
87–89
comfort/therapy and,
92–94
critics of, 65–67
disruptive technologies
and, 68–71
education and, 74–75,
90–92
health care and, 94–99
income and, 71–72
jobless future and, 67–68,
71–76
laws of robotics
and, 61–62
leisure and, 75–76

military applications,
99–106
morality and, 76–84
social benefits and, 72–74
space research and,
106–108
as threat to human species,
62–65
workplace pros/cons for,
84–86
humanoid robots, 28–38
artificial intelligence and,
30–31
in arts, 36–38
Cybernetic Zoo, 29
da Vinci and, 28–29
defined, 22
evolution of, 28
examples of, 31–33
first built, 22–23
humanlike characteristics
of, 29–30
Japan research of, 33–35
in literature, 35–36, 38

IBM Corporation, 30
ichor (life-giving fluid), 5
I-C-MARS Project, 151–156
Idaho Drone Law (2013),
226–228
IEEE council *versus* society,
181
IEEE Robotics &
Automation Society (IEEE
RAS), 181–184
immoral acts, 78–79

income in jobless future,
71–72
individual activity account
(IAA), 74
Industrial Revolution, 69
industrial robots, 24–28
annual supply of,
worldwide, 216
artificial intelligence and, 27
data on, 215–217
defined, 23
Devol and, 24
mobile robots, 26–27
operational stock of,
worldwide, 217
robot arms, 25–26
Silver arm, 27–28
Unimate as, 24–25
use of, by various
industries, 217
yearly shipments of, in
selected countries,
215–216
Institute for Ethics and
Emerging Technologies
(IEET), 177, 184–185
Institute of Electrical and
Electronics Engineers
(IEEE), 181
Intel, 131
International Federation of
Robotics (IFR), 185–187
Intuitive Surgical, Inc., 96
*Invention of Hugo Carbet,
The* (Selznick), 17
Isaacson, Walter, 194

Japan, humanoid robots and, 33–35
Japan Robot Association, 186
jobless future
    disruptive technologies and, 68–71
    education and leisure in, 74–76
    income in, 71–72
    robots and, 67–68
    social benefits in, 72–74
Johnson, David E., 132–136
Jordan, Angel, 205
Junod, François, 18

Kamen, Dean, 43, 109
Kasparov, Garry, 30–31
Katz, Lawrence, 76
Khizanat al-Hikma (The Treasury of Knowledge), 8
King, Ross, 137
Kismet (robot), 80
Kitab al-Hiyal (The Book of Ingenious Devices) (Banu Musa), 8
Kitab Fi Ma Rifat Al-hiyal Al-handasiyya (The Book of Ingenious Mechanical Devices) (Al-Jazari), 8–9
Kohlberg, Lawrence, 81
Kolb, Bliss, 18
Kraftwerk, 36
Krakovna, Viktoriya, 179
KUKA robotics, 138
Kurzweil, Ray, 64, 190–193
Kuyda, Eugenia, 159

laws of robotics
    AI singularity and, 63–65
    described, 61–62
    human species and, 62–65
    morality and, 82–83
Le Cat, Claude-Nichols, 212–213
LEGO Mindstorms EV3 system, 133
Lego robots, liquid-handling, 138
leisure in jobless future, 74–76
Leonardo da Vinci (Isaacson), 194
Lieh Tzu, 6
Littman, Michael, 66
Lokapannatti, 5–6
Lucid, 170
Luddite movement, 69–70
luddites, 69
Luka, 159
Lunokhod 1, 26

Machine Intelligence Research Institute (MIRI), 65, 196–198
machine tool, 19
macrostrategy, 177–178
manipulators, 106
Manum, Abdullah-al-, 8
Mason, Robert, 36
Mechanica (Heron), 7, 181
MedEthEx, 80–81
medical robots, 40–41
Méliès, Georges, 17

mens rea, 248
Metropolis (silent picture), 22, 23
Microsoft, 67, 72, 158
Miessner, Benjamin, 20
military robotics, 99–106
  delivery system, 102
  fire-fighting devices, 102–103
  mine detection/destruction systems, 101–102
  objections to, 103–106
  search and rescue robots, 103
  surveillance/reconnaissance, 100
  Tesla and, 99–100
  weapon-carrying devices, 101
mine detection/destruction systems, 101–102
Mine Kafon Drone, 102
Minsky, Marvin, 25–26, 198–200
mobile robots, 26–27
Model-Based Computer Vision (Brooks), 170
modern robotics, birth of, 20–23
Moore, George, 20
Moore, Gordon, 131
Moore's law, 131
moral agency, 80, 81
morality
  Asimov's laws of robotics and, 82–83

defined, 76
functional, 80
Kohlberg classifications of, 81
moral agency and, 80, 81
operational, 80
robots and, 76–84
"Mr. Eisenbrass" machine, |29
Murrin, Tom, 205
Musk, Elon, 63, 179, 200–201

National Robotics Engineering Center, 205
National Robotics Initiative, 2.0, 236–242
natural language processing, 158
Newstead, Keith, 18
North, Dug, 18
Norvig, Peter, 201
Numenta, 66
"Nutcracker, The," 17

Okpechi, Samuel C., 136–140
Open Roboethics Institute, 146
operational morality, 80
Opportunity (Martian land rover), 26–27
origami bot/robot, 98
Out Mathematical Universe (Tegmark), 179
Ozma of Oz (Baum), 35

PAX, 103
pedipulator, 26
pendant (teach box), 221
Perdix system, 101
"Petrouchka" (ballet), 17
"Phantom Doorman," 173
Philon of Byzantium, 6
Pneumatica (Ctesibius),
    10–11
Pneumatica (Heron), 7, 181
pneumatics, 7, 9
Pollock, Martha E., 201
"Preparing for the Future
    of Artificial Intelligence"
    (Committee on
    Technology), 243–244
Pressey, Sidney L., 90
Primitive Expounder
    (journal), 64
Programmable Universal
    Machine for
    Assembly, 25
Project Magneta, 38
proportionality principle in
    armed conflict, 104–105

Quesnay, François, 213

R. U. R. (play), 20–22, 36
Rancho Arm, 206
reconnaissance, 100
Reddy, Raj, 205
Rees, Martin, 62, 179, 180,
    200–202
Refell, A. H., 23, 32

Regulatory Robot (CPSC),
    244–245
rehabilitation, 96
rehabilitative medicine,
    robots in, 96–97
Replika, 159
Repp, Sierra, 140–144
Rest of the Robots, The
    (Asimov), 36
Rethink Robotics, 170–171
Richards, William H., 22–23
Rismani, Shalaleh, 144–147
Robonaut 2, 106–107
robot arms, 25–26, 27–28
robot camps, 108
robota, 130
Robotic Industries
    Association (RIA),
    202–205
robotic process automation
    (RPA), 87–89
robotic surgery, 228
robotically assisted surgery
    (RAS), 228–231
    benefits of, 228
    components of, 228–229
    described, 228
    health care providers,
        recommendations for,
        229–230
    patients, recommendations
        for, 230
robotics
    automata as precursor to,
        4–20

from automata to,
18–20
classifying robots, 38–43
courses/programs in,
108–109
defined, 4
European Parliament Draft
Report concerning,
233–236
evolution of, 23–24
humanoid, 22–23,
28–38
industrial, 23, 24–28
introduction to, 3–4
laws of, 61–62
military, 99–106
modern, birth of, 20–23
types of, 38–43
See also jobless future
Robotics Institute at Carnegie
Mellon University,
205–206
robotnik, 21
robots
agricultural, 86–87
AI singularity and, 63–65
business/finance, 87–89
classifying, 38–43
comfort/therapy, 92–94
criminal, 247–249
distribution data of, 218
domestic, 39–40
education and, 90–92
entertainment, 42–43
future of, in human society
(*see* human society,
future of robots in)

health care and, 94–99
as hobbies/educational
competitions, 43
Hooper essay on,
129–132
jobless future and, 67–68
medical, 40–41
military applications for,
99–106
moral, 76–84
Repp essay on, 140–144
search and rescue, 103
service, 41
sex, 92–94
social, 91–92
space research, 42,
106–108
weapon system, 41–42
workplace uses for, 84–86
"Robots, The" (song), 36
Rosheim, Mark, 193
Rossum's Universal Robots
(Čapek play), 130
Roxxxy (gynoid sex
robot), 92
"Runaround" (Asimov),
61–62, 163–164
Russell, Stuart, 201
Russian Association of
Robotics, 186

Sarkar, Anjali A., 147–151
Scheinman, Victor, 25,
206–208
Scorsese, Martin, 17
SeaBotix robots, 103
search and rescue robots, 103

self-driving vehicles, 231–232
Selznick, Brian, 17
service robots, 41
sex robots, 92–94
Shipboard Autonomous
  Firefighting Robot
  (SAFFiR), 102–103
Silver, David, 28
Silver arm, 27–28
simple machine, 9–10
Singularity Institute for
  Artificial Intelligence, 65,
  196
Singularity Is Near, The
  (Kurzweil), 64, 190
Skinner, B. F., 90
snake robot, 141
social benefits
  defined, 72–73
  in jobless future, 72–74
social robot, 91–92
  Sarkar essay on, 147–151
  WOZ-controlled, 150–151
Society of Mind theory, 198
Society of Model
  Engineers, 23
Solo (Mason), 36
Southwestern Indian
  Polytechnic Institute
  (SIPI), 153–155
Space General
  corporation, 26
space research robots, 42,
  106–108
special needs students,
  educational robots and,
  91–92

Spirit (Martian land rover),
  26–27
Spooner, Paul, 18
Stanford Arm, 206
STEM (science, technology,
  engineering, and
  mathematics), 91,
  132–133, 142
  I-C-MARS Project and,
  151–156
Stepford Wives, The
  (film), 37
Stochastic Neural Analog
  Reinforcement Computer,
  198
strong AI, 65
supervised training of bots,
  158
surgical robot, 40–41,
  95–96. *See also* Da Vinci
  Surgical System
surveillance, 100

Tallinn, Jaan, 178
Talos, 5, 35
Tambourine (Tabor) Player,
  211
Tangiblek, 90–91
*Taylor v. Intuitive Surgical,
  Inc.,* 249–251
technoprogressive
  technologies, 184
Tegmark, Max, 178–179
Teledyne Marine, 103
telemedical system, 98
Tesla, Nikola, 20,
  99–100

therapy robots, 92–94
Three Laws of Robots
  (Asimov), 163–164
3M, 186
top-down robotics, 169
Transitions Research
  Corporation, 176
treebot, 86–87
Truitt, E. R., 14–15
Turing, Alan, 31, 208–211
Turing test, 208–209

Unimate, 24–25
Unimation Corporation,
  25, 173, 175
United Cinephone, 172
universal basic income
  (UBC) program, 71
unsupervised training of
  bots, 158

Vadiee, Nader, 151–156
Vaucanson, Jacques de, 16
Verrocchio, 194
Virgil, 12
Virginia Robot Delivery
  Laws, 2017, 246–247

WABOT-1 (Japan humanoid
  robot), 33–34
WABOT-2 (Japan humanoid
  robot), 34
walking truck, 26
Wallach, Wendell, 77, 80–81
Walt Disney Company, 42
Wang Yangming, 18
weapon system robots,
  41–42
weapon-carrying devices, 101
Weizenbaum, Joseph, 157
Westinghouse Electric
  and Manufacturing
  Corporation,
  32–33, 205
Westworld (film), 37
Wilkinson, John, 19
WOZ-controlled social
  robots, 150–151
Wozniak, Steve, 63
Wright, S. Fowler, 36

Xi Jinping, 18

Yearsley, Liesl, 159

Zapata, Carlos, 18
Zimmerman, Erin, 156–161

About the Author

David E. Newton holds an associate's degree in science from Grand Rapids (Michigan) Junior College, a BA in chemistry (with high distinction), an MA in education from the University of Michigan, and an EdD in science education from Harvard University. He is the author of more than 400 textbooks, encyclopedias, resource books, research manuals, laboratory manuals, trade books, and other educational materials. He taught mathematics, chemistry, and physical science in Grand Rapids, Michigan, for 13 years; was professor of chemistry and physics at Salem State College in Massachusetts for 15 years; and was adjunct professor in the College of Professional Studies at the University of San Francisco for 10 years.

The author's previous books for ABC-CLIO include *Global Warming* (1993), *Gay and Lesbian Rights—A Resource Handbook* (1994, 2009), *The Ozone Dilemma* (1995), *Violence and the Mass Media* (1996), *Environmental Justice* (1996, 2009), *Encyclopedia of Cryptology* (1997), *Social Issues in Science and Technology: An Encyclopedia* (1999), *DNA Technology* (2009, 2016), *Sexual Health* (2010), *The Animal Experimentation Debate* (2013), *Marijuana* (2013, 2017), *World Energy Crisis* (2013), *Steroids and Doping in Sports* (2014, 2018), *GMO Food* (2014), *Science and Political Controversy* (2014), *Wind Energy* (2015), *Fracking* (2015), *Solar Energy* (2015), *Youth Substance Abuse* (2016), *Global Water Crisis* (2016), *Youth Drug Abuse* (2016), *Same-Sex Marriage* (2011, 2016), *Sex and Gender* (2017), and *Sexually Transmitted Diseases* (2018). His other recent books include *Physics: Oryx Frontiers of Science Series*

(2000); *Sick!* (four volumes) (2000); *Science, Technology, and Society: The Impact of Science in the 19th Century* (two volumes, 2001); *Encyclopedia of Fire* (2002); *Molecular Nanotechnology: Oryx Frontiers of Science Series* (2002); *Encyclopedia of Water* (2003); *Encyclopedia of Air* (2004); *The New Chemistry* (six volumes, 2007); *Nuclear Power* (2005); *Stem Cell Research* (2006); *Latinos in the Sciences, Math, and Professions* (2007); and *DNA Evidence and Forensic Science* (2008). He has also been an updating and consulting editor on a number of books and reference works, including *Chemical Compounds* (2005), *Chemical Elements* (2006), *Encyclopedia of Endangered Species* (2006), *World of Mathematics* (2006), *World of Chemistry* (2006), *World of Health* (2006), *UXL Encyclopedia of Science* (2007), *Alternative Medicine* (2008), *Grzimek's Animal Life Encyclopedia* (2009), *Community Health* (2009), *Genetic Medicine* (2009), *The Gale Encyclopedia of Medicine* (2010–2011), *The Gale Encyclopedia of Alternative Medicine* (2013), *Discoveries in Modern Science: Exploration, Invention, and Technology* (2013–2014), and *Science in Context* (2013–2014).